Head Strong

Head Strong

How Psychology is Revolutionizing War

Michael D. Matthews

OXFORD
UNIVERSITY PRESS

Oxford University Press is a department of the University of Oxford.
It furthers the University's objective of excellence in research, scholarship,
and education by publishing worldwide.

Oxford   New York
Auckland   Cape Town   Dar es Salaam   Hong Kong   Karachi
Kuala Lumpur   Madrid   Melbourne   Mexico City   Nairobi
New Delhi   Shanghai   Taipei   Toronto

With offices in
Argentina   Austria   Brazil   Chile   Czech Republic   France   Greece
Guatemala   Hungary   Italy   Japan   Poland   Portugal   Singapore
South Korea   Switzerland   Thailand   Turkey   Ukraine   Vietnam

Oxford is a registered trademark of Oxford University Press
in the UK and certain other countries.

Published in the United States of America by
Oxford University Press
198 Madison Avenue, New York, NY 10016

© Oxford University Press 2014

The views expressed in this book are those of the author and do not reflect the
official policy or position of the Department of the Army,
Department of Defense, or the US Government.

Library of Congress Cataloging-in-Publication Data
Matthews, Michael D.
Head strong : how psychology is revolutionizing war / Michael D. Matthews.
pages cm
Includes bibliographical references and index.
ISBN 978-0-19-991617-7 (acid-free paper)
1. War—Psychological aspects.   2. Psychology, Military.   3. Combat—Psychological
aspects.   4. United States—Armed Forces—Medical care.   5. Psychological warfare—United
States.   6. Soldiers—Mental health—United States.   I. Title.
U22.3.M38 2013
355.001'9—dc23
2013026357

1 3 5 7 9 8 6 4 2
Printed in the United States of America
on acid-free paper

This book is dedicated to the men and women of the United States Armed Forces who have served so bravely in our nation's wars, past and present, and to those who will follow in your footsteps in coming years. You represent the best that our country has to offer, and I thank you for your selfless service and sacrifices.

CONTENTS

LIST OF ILLUSTRATIONS

FIGURES

TABLE

FOREWORD

It is ironic that we still see books, doctrine, video, and other references to psychological warfare as if it were some special categorization, as if psychology and warfare were somehow separate until joined together through scholarship or strategy. In point of fact, psychological warfare is the only kind of warfare there is. Warfare is a test of wills, and will is a psychological construct. It is exciting to see a new book about the psychology of war, because only such a book has a reasonable chance at getting to our understanding of war at its core—in the human dimension. If war is, as Carl von Clausewitz so elegantly stated, "politics by other means," then those politics and means are inherently psychological. War only exists in the human dimension.

Despite the nature of war as psychological, traditional fascination with the mechanical implements of war remains deeply embedded in military culture. The throwbacks are pervasive. Service academies continue to wrap their curricula around engineering competencies, just as they did in 1802 when Thomas Jefferson established the US Military Academy, to train engineers and artillerists to push the nation's expansion westward. Congressmen speak of a 600-ship navy and talk in terms of numbers of fighter jets, as if industrial age attrition warfare is still a possibility. Drones, the newest lethal technology, provide tremendous precision strike capability while allowing pilots to have dinner with the family in the club. Recently, a researcher with the Defense Advanced Research Projects Agency (DARPA) spoke of new cyberwar technology, stating that it is "like crack for generals." A *Wired* article on the project boasted with its title "This Pentagon Project Makes Cyber Warfare as Easy as *Angry Birds*."[1] Such claims would lead one to believe that we are incapable of losing wars, and that the military is in good shape.

In actuality, our current military leadership is impaled on the psychology of it all. As of this writing, 72 acts of sexual misconduct occur in the active military each day. Twenty-two suicides among veterans occur every day. Congress is responding to popular perceptions and emotional outrage over a proposed military award for drone pilots that would take precedence over valorous awards for fighting on the ground. Our air force has suspended 17 nuclear missile launch officers because of substandard duty performance, much of it related to weak leadership, poor morale, and feelings of

---

1. See www.wired.com/dangerroom/2013/05/pentagon-cyberwar-angry-birds/2/.

worthlessness in a dead-end career path. The *New York Times* recently announced that the Taliban have begun a process of peace negotiations with the Karzai government in Afghanistan—a most delicate psychological balance. Clearly the art of generalship is dominated by the science of psychology. Military challenges are not in inventing the best machines. The challenge is in the manning, and that is principally informed by psychology. As I read this book, I viewed it as a natural outgrowth of Mike Matthews's extensive consultancy to the Chief of Staff, Army, the Vice Chief of Staff, Army, and a host of other general officers who are faced with military problems best informed by psychological science.

No author better connects the science of psychology with military endeavors than Mike Matthews. *Head Strong: How Psychology is Revolutionizing War* will take the everyday reader into the inner processes and dynamics of how to make a 1.2 million-person organization find its members, train them in what do, shape their thinking in constructive ways, and apply elements of science to shape effectiveness from the battlefield to the board room. What's more, despite the immense psychological challenges facing our military, Matthews's scholarly tradition and body of work is one of positive psychology, bringing an upbeat character to what might otherwise be a deeply critical, pessimistic writing.

Looking back 35 years on my own military career, I wish I would have had this book and benefitted from its insights. Much of my career in the army was spent trying to innovate my way around and through a tightly coupled bureaucracy. Innovation occurs for all of us when we are able to take a unique perspective on a problem or issue. Innovators see things differently than others, and in the seeing, can chart a different way ahead. As you read this book, you will be led to see issues of personnel, of effectiveness, and of management in ways different than when you began reading, and certainly different than your peers around you. The ability to innovate inside of bureaucracies is priceless. *Head Strong* will help you do that, and should be a top-shelf read for everyone in the military.

My greatest delight in reading this book however, is not the connectedness of psychology with military matters. Instead, it's the deeply insightful structure of the book and the broadening parallels to business and life. For example, Matthews starts by articulating the relationship between psychology, strategy, and the art of war; moves through the principal human resource management functions of matching the right person with the right job, to onboarding and acculturating employees; to building cognitive systems that ensure competitive advantage, and leading virtual teams. This list of topics reads like a tour de force of industrial/organizational psychology as applied in business. It is likewise a practical how-to for the modern C-suite executive. Psychological systems thinking is one of the more pragmatic ways of viewing individuals in organizations, as well as the organizations themselves. That Matthews applies this knowledge to military people, topics, and situations does not make this only a military book, or of sole interest to military people. Readers will find, as I did, a tremendously broad applicability of this book to people—especially leaders—in all walks of life.

The chapter on leading in combat provides a specific example of the broad applicability of *Head Strong*. Yes, combat is an inherently military topic. Matthews reviews current contemporary science and thinking about interpersonal influence in dangerous contexts, citing respected and well-published psychologists like Pat Sweeney, Paul Bartone, and Steve Zaccaro. Yet each principle proposed in this chapter is equally applicable in elite business and other high-risk endeavors where livelihoods are on the line. Students of crisis need not plow through case studies of companies on the ropes and their struggling executive leadership, the majority of whom never sought to be challenged by a crisis. Instead, Matthews's chapter on combat reveals crisis lessons taught by people who, as a matter of their profession, run toward crisis as opposed to running away from it. Most of us agree that we would prefer to learn from professionals than amateurs. *Head Strong* gives us such an opportunity.

Our country is made safer and better because of thought leaders like Mike Matthews and the applied analytical thinking characterized by *Head Strong*. Readers—whether in the military or in everyday life—will grow to understand more fully how people connect with national defense. At the same time, we all can learn much about our own organizations by seeing the science of psychology inform the strategy, operations, and improvement of one of our largest and most resilient national institutions—our military. I am a social psychologist by training, served as a military leader by both passion and necessity, and now teach at a business school. This book challenged and influenced my thinking in each dimension of my life, as I am sure it will yours.

Thomas A. Kolditz, PhD
Brigadier General, US Army (Ret.)
Professor, Yale School of Management

PREFACE

The idea for this book came to me at an international military psychology meeting in Delhi, India, in March of 2011. It suddenly struck me that military psychologists from around the world were all interested in the same topics, and that the very success of military operations in the twenty-first century depends more than ever before on psychology. I sensed a hunger among the audience for knowledge of how psychology will continue to influence future wars, and decided I would write a book that explores this question. In retrospect, I am not sure whether to attribute my idea to inspiration or jet lag. In truth, both probably played a role.

I have a great passion for this topic. Military psychology is fascinating because it studies human behavior under extreme conditions. Life-and-death stress, sleep deprivation, separation from home and family, the uncertainty of dealing with people from cultures entirely unlike our own, living the disciplined life so necessary to the military—these conditions magnify the importance of psychology and provide a venue for studying human behavior unlike any other. On top of this add revolutionary advances in technology, and you get the recipe for psychology as a game-changer in twenty-first-century war.

In this book I use the word "soldier" in a generic sense to refer to any person who serves in uniformed military service of their country. It is not my intention to slight the service of the men and women who serve so valiantly in our navy, marine corps, and air force. They, along with their army counterparts, are all soldiers in the sense of serving in the profession of arms. Whether they serve on a ship, an airplane, or a tank is secondary to the fact that they make so many sacrifices in support of national defense. Accordingly, in this book I refer to these warriors as soldiers, and will use terms like sailor, marine, or airman when comments are specifically relevant to a particular branch of service.

This book also assumes a United States centric viewpoint. Unless I say otherwise, I am referring to US armed forces. That said, virtually everything I say in this book should be equally relevant to the armed forces of other nations. Over the years I have collaborated with military psychologists from Australia, Canada, Singapore, Norway, France, Switzerland, India, and Germany. At military psychology conferences, I have listened to psychologists from many other nations address issues deemed vital to the military of their own countries. Without exception, the questions that military psychologists address are universal. The degree of overlap in both research and application of military psychology across nations never ceases to astound me.

I have intentionally written this book in a voice aimed at the general public as well as for military psychologists. When writing for other psychologists, it is easy to fall into arcane jargon. I hope that anyone with an interest in the study of war—and in particular the human element in war—will find this book interesting. I would be delighted if this book inspires undergraduate or graduate students to enter the profession of military psychology. Toward this end, I have tried to minimize extensive reference to very specialized scientific articles, and chose to use end notes rather than the parenthetical citations more commonly used in scientific writing. It is a delicate balance between making these chapters readable for the general public, yet still informative for psychologists and other behavioral and social scientists.

Writing about the future is always a chancy affair. It is easier to "predict" the past than the future, to be sure. The military is often blamed for being ready to fight the last war, not the next one. I have set my sights on the timeframe of 2030 to 2050. I am sure I will have a good laugh at myself if I am lucky enough to be around then and can see how my prognostications fail. Some of my predictions simply won't pan out. And things that I can't imagine now will turn out to be vital. But one thing I am sure of is this: Psychology will be of fundamental importance to who wins and who loses wars of the future. Militaries that embrace the role of psychology will do a better job of supporting the vital national interests of their nation than those who fail to bring psychology to bear on military matters.

I might not have written this book if it *only* mattered to the military. But knowledge generated by military psychologists extrapolates directly to the civilian world. And this knowledge serves to improve the quality of life for everyone, not just soldiers. "The military," after all, is comprised of ordinary people who are called upon to do extraordinary things, and what we learn from military psychology translates easily and naturally to the civilian world. Future developments in military psychology will improve employee selection and training, help us make emerging technologies more effective and useable, and enable people to develop the skills they need to live happy, productive, and well-adjusted lives. This is not conjecture; it is fact. As you will see, the wars of the twentieth century drove developments in psychology that directly impact our lives today. Wars of the twenty-first century will fuel even more exciting developments that will change our lives for the better.

Other books on military psychology tend to focus on stress, pathology, and clinical psychology. There is a strong dose of these topics in this book, but I have endeavored to look at all areas of military psychology, not just areas of clinical importance. I write about selection, training, resilience, human factors engineering, cognitive and biological psychology, leadership, ethics, and peace. As reflected in these topics, the field is so broad that to call oneself a military psychologist does not provide much of a clue as to exactly what type of psychologist one is. Psychologists study human behavior. Normal, abnormal, and in-between, you will find it all here.

If I have done a good job writing this book, you will want to learn more. For more in-depth and technical information on military psychology, past and present, I'd recommend the *Oxford Handbook of Military Psychology*, published in 2012. Dr. Janice Laurence and I edited this book to capture the current state-of-the-art of military psychology. The 27 chapters touch on all aspects of military psychology. The chapter authors are today's leading military psychologists. The handbook may be a bit more palatable to psychologists than the general reader, but I think anyone with a genuine interest in the field will find it of value.

You may also find it interesting, having read this book, to be on the lookout for news stories that pertain to military psychology. Employ your social media tools. Psychologists may wish to join the APA Division 19–Military Psychology group on Facebook. Hot topics in military psychology—stress, suicide, and training, for example—may be trending on Twitter or other media. If you are a student, take a course in psychology and select a topic relevant to the military for your semester term paper. Don't just focus your research on stress and resilience; instead, apply what you learn about stress and resilience to the military context. And I love talking about these issues. Send me an e-mail at lm6270@usma.edu, and I'll answer it. Over the years, I have talked with dozens of high school and college students who have taken an interest in this field. I'll reply—really!

If psychology will continue to be a difference-maker in wars of the twenty-first century, it is also true that psychological knowledge should be leveraged to promote peace. A systematic and well-funded agenda aimed at developing a better understanding of the scientific basis of peaceful behavior at both the group and individual level could go a long ways toward making the world a safer place for everyone. Both war and peace are complex social phenomena, but both are amenable to scientific analysis. Our Department of Defense does a good job funding military psychology. I'd like to see equal funding for psychologists studying peace.

Military psychology represents an interesting story. Beginning in World War I and continuing to the present time, war has driven paradigm-changing advances in psychology. The wars yet to come in the twenty-first century will do the same. It will be interesting to watch the story unfold.

ACKNOWLEDGMENTS

You learn from serving in the military that one's personal achievements and accomplishments are inextricably tied to the support and efforts of others. To be sure, I sat in front of my computer for hundreds of hours writing this book, but the knowledge, skills, and ideas I needed to complete this work come from the psychologists, students, colleagues, and soldiers I have known over the past 33 years.

First and foremost, I thank my parents, Delbert and Maurine Matthews. Dad was a World War II vet, both were teenagers during the Great Depression, and they both wanted their children to be the first generation in their families to receive a higher education. It is only now, when I am older and they are gone, that I really understand the true and lasting value of the sacrifices they made to ensure that my brothers and I received our college degrees. Thanks, Mom and Dad, for getting me off to a good start in life.

I thank my undergraduate psychology mentor, Dr. Victor M. Agruso Jr., for both inspiring me to become a psychologist and instilling in me the value of using science to answer questions about human behavior. I was the grateful recipient of a first-class undergraduate education in psychology at Drury College (now University), thanks to Victor's mentorship and support. I wanted to be just like him when I grew up. I failed in that goal, but hope that he is satisfied with how I turned out.

After graduating from Drury, I entered the masters program in psychology at Hollins College. This small college had a terrific predoctoral masters program in those days, and I learned a great deal from my professors there: Lowell Wine, Paul J. Woods, Randall Flory, Ronald Webster, and F. J. McGuigan. They were all notable researchers who were also great teachers.

I obtained my doctorate in experimental psychology at the State University of New York at Binghamton. My dissertation advisor, Dr. Harold (Hal) Babb, had served as a military policeman in World War II. I had already joined the air force before I got around to defending my dissertation. I remember Hal grinning from ear to ear and predicting I would spend my career in the air force. He was wrong (ah, the danger of predicting the future!), but I have spent my career as a military psychologist. Hal, like Victor at Drury and the Hollins professors, molded my early worldview of psychology. I stand on their collective shoulders.

From my time in the air force to the present, I have been strongly influenced by many military psychologists, both uniformed and civilian. At the Air Force Human Resources Laboratory, I recognize the influence of Bill Alley, Gene Berry, Joe Ward, Ray Christal, and Bob Gagne. At the Air Force Academy my office mate, Bill Cummings, kept me both intellectually stimulated and entertained. (The cadets will never forget your chimpanzee act!) I got my first real dose of army psychology, oddly enough, at the US Air Force Academy (USAFA). Lieutenant Colonel Rufus Sessions, an army medical services corps psychologist, was on an exchange assignment in my department at USAFA. He and I collaborated on several major research projects, team-taught the biopsychology course, and drank our share of beer. Rufus died at a young age nearly 10 years ago. I celebrate his memory every New Year's Day, when I use his recipe to make Georgia-style black-eyed peas and salt pork. He was a great scientist and friend.

Many other military psychologists have inspired me. I owe a debt of gratitude to my former colleagues at the US Army Research Institute's Fort Benning office: thank you Scott Graham, Jean Dyer, Marnie Salter, Bob Pleban, and Rhett Graves. Lots of folks from other organizations have shaped my thinking, and I'd like to single out Mica Endsley, Laura Strater, Rich Wampler, Janice Laurence, Paul Bartone, Armando Estrada, Mike Rumsey, Steve Goldberg, and Nita Miller Shattuck. In different ways, you pushed me in good directions. Brigadier General Rhonda Cornum, US Army (Ret.), is not a psychologist (I forgive her), but she made a significant contribution to military psychology by "standing up" the army's resilience program, Comprehensive Soldier Fitness. She is a true American hero and I am proud to count her among my friends.

Major General (Ret.) Robert H. Scales is a strong advocate for the importance of psychology in the twenty-first-century military. A member of the West Point Class of 1966, Bob's combat experiences in Vietnam shaped his later ideas about psychology and war. I met him soon after he retired from active duty, where his last job was Commandant of the US Army War College. I was fortunate to assist Bob and other like-minded people in their quest to improve military selection, training, leadership, and combat effectiveness through psychology and the other behavioral sciences. You would not be reading this book without his influence and mentorship.

There are a number of military psychologists from other nations that I would like to thank for their inspiration and collegiality. Thanks to the following: Jarle Eid, Bjorn Helge Johnsen, Ole Boe, Hubert Annen, Uzi Ben Shalom, Tin French, and N. P. Singh and his colleagues with the Indian Military Psychology group in Delhi.

My colleagues at West Point deserve very special recognition. In 2000 Larry Shattuck saw fit to hire me to teach in the Engineering Psychology Program. He continues to be a great friend and collaborator from his post-army position on the faculty of the Naval Postgraduate School. Larry never fails to remind me that January in

Monterey is much nicer than January at West Point. And I remind him that adversity builds character!

I have had the good fortune to work for three terrific department heads at West Point. Brigadier General (Ret.) Casey Brower, now at the Virginia Military Institute, endorsed hiring me and helped me get a foothold in the Department of Behavioral Sciences and Leadership. His successor, Brigadier General (Ret.) Tom Kolditz, now a professor of leadership at Yale University, gave me the freedom to study interesting things and always had my back. Our current department head, Colonel Bernard Banks, picked right up when Tom retired and I thank him for his unflinching support.

West Point sociologist Morten Ender is a fountain of great ideas. The Biannual Survey of Students (BASS), a now 10-year study of the social attitudes of academy cadets, ROTC cadets, and nonmilitary affiliated college students began with a hallway conversation. Morten likes the couch I have in my office and comes to visit almost every day. I suspect that several journal articles and books will someday be directly attributable to ideas generated from that couch!

I also want to call out special thanks to Colonel (Ret.) Patrick Sweeney. Pat taught in our department and is both a friend and collaborator. We taught a course together in combat leadership and based on the lesson plans from that course developed an edited volume, *Leadership in Dangerous Situations* (Naval Institute Press, 2011). Pat is a terrific leader and I am enjoying watching his post-army career flourish. Thanks Pat, for finding so many engaging topics on which to collaborate. I look forward to many more.

Thank you, Tim O'Neill, for standing up the Engineering Psychology program back in the 1980s. Donna Brazil, thanks for your support of army resilience training. Don Campbell, your ideas about in extremis leadership are cutting edge; thanks for sharing. Dan Smith, your discourses on psychology, cadets, and the state of the world have made the miles melt away as we train for our half-marathons. James Merlo, you are the most energetic intellect I have ever met; keep those great ideas flowing! Jim Ness—you are the most broadly educated uniformed psychologist I have ever met. I value your sense of humor and your devotion to physical fitness. Your pull-up record will never be broken. Dennis Kelly, you are my most trusted collaborator. If we ever publish all of the data we have collected, we will be famous. Lolita Burrell, Kathy Campbell, Ericka Rovira, Margie Carroll—and all the other faculty—you are great colleagues and I look forward to seeing you each and every day. Thanks to the military faculty. You have all been to war and served with honor and distinction. It is humbling to teach side by side with you.

Thank you to the cadets I teach. Your energy and optimism keep me on my toes at all times. I owe it to you to teach you a psychology that will make you better leaders. If I am lucky enough to live a long life, I expect to be reading about your superlative accomplishments as the years unfold. Maybe one of you will be elected president someday. If so, you won't be the first West Point graduate to do so. And you face a tougher job than you can know. Too many of you have been killed or wounded in Iraq or Afghanistan. Work hard and flourish, but be safe.

Much of my focus in military psychology relates to the emerging field of positive psychology. I must single out Martin Seligman of the University of Pennsylvania for special thanks. Our careers mirror each other's in some important respects. Both of us began our research careers in animal learning, specifically aversive conditioning. Marty, as almost everyone knows, gained fame for his theory of learned helplessness. I, as virtually nobody knows, began by conducting research on self-punitive behavior in rats. We both soon developed interests in more general psychology. Our paths eventually crossed at West Point in 2004, when Marty visited our department. We struck up a conversation about the relevance of positive psychology to the military context. In 2005 Marty selected me to be part of the Medici II Positive Psychology group, which met at Penn that summer. The ideas we developed during the Medici laid the groundwork for numerous research projects and, more importantly, manifested themselves in the US Army's Comprehensive Soldier Fitness program, perhaps the largest application of psychological principles to a major social institution in the history of the discipline.

I soaked up ideas from other psychologists during the Medici. Mike Csikszentmihalyi, Ed Diener, Nansook Park, and George Valliant—along with another dozen or so graduate students and psychologists—explored positive psychology and its relevance to the human condition. From these discussions I extracted a good number of ideas that seemed especially germane to the military.

I have not yet mentioned Christopher Peterson. Chris was a mainstay at the Medici, but his influence on me continues to this day. Tragically, Chris died unexpectedly in the fall of 2012. With his friend and research partner Nansook Park, we (along with Pat Sweeney) were conducting some truly unique research on the effects of combat deployments on subsequent psychological development. Chris was a great human being. He was one of the most gifted scholars I have ever met, yet was modest, cared more about others than himself, and was a true friend to all who knew him. Chris's influence will live on through his impact on the field of psychology and in the hearts of all who knew him.

Dr. Angela Duckworth, University of Pennsylvania, also deserves special thanks. Angela invented the concept of grit, a measure of the passionate pursuit of long-term goals. She collected part of her dissertation data at West Point, and was kind enough to extend coauthorship to me on a subsequent article on grit published in the *Journal of Personality and Social Psychology* (June 2007). If I were to select one young psychologist to follow for the next 25 years, it would be Angela. Her research addresses important issues and has wide impact. Besides that, Angela is a wonderful person. You can consider me the founding member of the Angela Duckworth fan club. She is going to do great things in psychology. Thanks for letting me be a part of your effort!

I'd like to convey special gratitude to retired army colonel Jill Chambers and her husband, Michael Peterson. Jill was instrumental in helping the army understand how to go about improving its approach to mental health, an effort that ultimately resulted in the establishment of the Comprehensive Soldier Fitness program. Michael uses his talents as an artist and musician to inspire untold thousands of active duty military members, their families, and veterans each year. Together, they have helped me expand my focus in military psychology. Thanks, Jill and Michael, for all you do and will continue to do in support of our military and veterans.

The folks at Oxford University Press have been absolutely delightful to work with. Thank you, Abby Gross, for your confidence in both me and this project. You and the editorial team, especially Suzanne Walker, are magicians! You have helped me take a raw idea and bring it to a polished fruition. I'd recommend OUP to anyone considering writing a book.

Finally I want to thank my wife, Gretchen Bain Matthews, for everything she has done to support this project. Gretchen is a Pomona College graduate (go Sagehens!) and English major. To my very great benefit, she has worked as a technical writer and editor since graduating from college. So besides offering moral support (much appreciated!) she has helped edit and clean up drafts of this book and honcho the technical aspects of the writing process. When it comes to editing, there is nothing she can't do. This includes being the consummate diplomat, skillfully correcting my writing without threatening my ego. And it gets better. Gretchen's mother, Barbara Bain (also a Sagehen English major), proofed each chapter. She said they were all "wonderful." I took that as an entirely unbiased critique. Not wanting to overburden their kind and helpful dispositions, I have promised both Gretchen and Barbara that I won't undertake another project like this again soon. After all, the next edition should not be due until about 2050.

Head Strong

1

PSYCHOLOGICAL SCIENCE
AND THE ART OF WAR

There's a graveyard in northern France where all the dead boys from D-Day are buried. The white crosses reach from one horizon to the other. I remember looking it over and thinking it was a forest of graves. But the rows were like this, dizzying, diagonal, perfectly straight, so after all it wasn't a forest but an orchard of graves. Nothing to do with nature, unless you count human nature.

Barbara Kingsolver[1]

The United States Military Academy, popularly referred to as West Point due to its location on a rugged piece of land jutting out from the west side of the Hudson River, is the longest continually occupied military reservation in the United States. West Point was of strategic importance to the colonies during the Revolutionary War, over two decades before the academy itself was founded in 1802. As a long-time professor of psychology at West Point, I often find myself giving walking tours of the grounds to visitors. Among the points of interest that I always show visitors is Trophy Point. Trophy Point sits high on a bluff looking north along the Hudson River, at the point where it narrows and bends in a sharp curve to the east before straightening and widening for its final 60-mile flow into New York City. Keeping British warships from safely navigating north of West Point was vital to the colonists, because if the British could control the river, they could split the colonies in half, yielding a huge strategic advantage.

To control the river, the colonists built artillery redoubts on each bank. Because the river is so narrow at this point, the British ships were within easy range of the guns. But the colonists also devised a brilliant plan to further deter the British fleet from navigating north of West Point. They forged large iron chain links (a sample can be found today on display at Trophy Point; see Figure 1.1) into a single long chain known as the Great Chain, placed the complete chain on timbers, and then floated the chain across the entire width of the river. One can only imagine the reaction of a British naval captain as he executed the challenging task of maneuvering a sailing ship up the river against both current and tides, only to come around the bend in the river and see the extensive gun emplacements and a massive chain barring further progress up the river.

Not a single captain attempted to ram the chain and sail his ship north from this point. As a result, the colonists maintained control of this vital resupply route, a critical victory in an era when rivers were the main mode of commerce in North America.

It is easy to view this military outcome as stemming from a remarkable engineering accomplishment. The Hudson is over a quarter mile wide where the chain was floated across the river. The Great Chain was approximately 500 yards long and consisted of two-foot-long, 2.25-inch thick iron links joined together and attached to logs. Stringing the chain across a river of this width against the forces of current, tide, and wind was a remarkable engineering accomplishment. My West Point colleagues in the engineering disciplines no doubt tout the engineering genius of our revolutionary forefathers as they give tours to their own visitors. Interestingly, however, modern engineers have calculated that the great chain could not have withstood the force of a British warship ramming it. In other words, the chain became a psychological barrier, in effect a form of eighteenth-century "psyops" (contemporary military shorthand for psychological operations). The chain accomplished the goal of blocking the river without the colonists relying on cannon fire—something the dominating British navy would certainly not have shied away from in and of itself.

I begin this book with the example of the Great Chain because it illustrates how psychology—combined with engineering and technology—has influenced the

FIGURE 1.1 West Point's Trophy Point, with links of the Great Chain in the foreground. (Photo courtesy of author.)

outcome of war throughout history. As every student of psychology knows, psychology as a distinct field of study and practice did not emerge until the late nineteenth century. But, as Herman Ebbinghaus is widely quoted as saying, "psychology has a long past but only a short history."[2] The importance of understanding human nature and how it affects and is affected by war reaches back through all of recorded human history. In *Achilles in Vietnam*, Jonathan Shay compares the human impact of war in Homer's *Iliad* with modern warfare, specifically in Vietnam, an account that is strikingly moving.[3] Perhaps the greatest early account of war is Sun Tzu's *The Art of War*. Written over 2,500 years ago, it is still required reading in military colleges. Sun Tzu tells us, "If you know the enemy and know yourself, you need not fear the result of a hundred battles."[4] This brilliant military strategist clearly recognized the fundamental importance of understanding human nature—both one's own and that of others—in success in war.

HISTORY OF PSYCHOLOGY AND WAR

Military psychology has been important to psychologists in the United States from the early days of the discipline. The American Psychological Association (APA) included a division of military psychology among its first group of specialized divisions. Now known as Division 19, the Society for Military Psychology, it represents over 500 psychologists who devote all or much of their scientific and professional energies to questions of importance to the military. These interests run the gamut from traditional clinical psychology and how it can be used to treat psychological casualties of war, to basic behavioral and cognitive neuroscience. I had the privilege of serving as president of Division 19 in 2007 and 2008, and was struck with the variety of contributions that psychologists are making to support our soldiers in the current wars.

The role of military psychology as a recognized scientific discipline begins in World War I. This was the first war fought with modern technologies. Tanks and airplanes were introduced to the battlefield, along with more powerful and accurate rifles, machine guns, and artillery. For the first time weapons of mass destruction—in this case chemical weapons—were unleashed. The internal combustion engine replaced the horse as the primary means of moving equipment, supplies, and troops to and from the battlefield. Despite these developments, nineteenth-century doctrine, tactics, techniques, and procedures were used to engage the enemy. Large armies faced off from each other, like opposing football teams, and employed their weapons with drastically lethal effects. Infantry charges designed for the era of the musket now resulted in almost certain death. Armies retreated to their respective trenches, awaiting the order for their next suicidal charge.

This mismatch between technology and doctrine was compounded by the ever-increasing complexity of the weapons and other systems used in battle. A Civil War soldier did not need to know much technology except how to load, fire, and care for his musket. The musket itself was a simple firearm with relatively few moving parts, and other than keeping it clean, it required little maintenance. But the infantry soldier in World War I needed to know how to employ relatively complicated bolt-action rifles, semiautomatic handguns, and fully automatic machine guns. Navy ships carried massive and sophisticated weapons, and submarines were introduced to warfare. Perhaps the most electrifying development was the use of aircraft as a means of delivering bullets and bombs to the enemy.

War had become sufficiently complicated that soldiers had to be selected based on special aptitudes and abilities and trained to operate their increasingly complicated weapons. In 1917, Robert Yerkes, the president of the APA, recognized the role that psychology could play in helping the US Army improve its selection and training procedures. Yerkes issued a "call to arms" to the members of the APA.[5] Amazingly, half of the 300 members volunteered to serve in one manner or another, and ultimately several hundred psychologists assisted the army, mostly in roles that facilitated the selection, assignment, and training of soldiers.

Perhaps the signature contribution of psychology during World War I was the development of the Army Alpha and Army Beta intelligence tests, a task completed under the supervision of Dr. Yerkes himself. This was no small undertaking, as it involved testing over 2 million soldiers. In an era long before the development of computers, the amount of raw effort that went into this task is almost incomprehensible. Much of the credit for the modern intelligence test can be attributed to this work.

World War II also spurred rapid developments in psychology. Selection tests continued to be revised and improved, proficiency tests were developed for scores of different military occupational specialties, and human factors engineering emerged as a subdiscipline of psychology. As the machines of war came to challenge the limits of the ability of humans to successfully operate them, more care had to be taken in the design of weapons and their supporting systems. One example is the increased speed and complexity of airplanes. World War I warplanes were slow and were navigated by sight. By World War II airplanes could fly several hundred miles per hour, carried massive amounts of sophisticated weapons, had complicated communications equipment onboard, and required exceedingly complex instrumentation, as can be seen in the photo of the cockpit of a World War II-era B-24 bomber (Figure 1.2). Displays and controls had to be laid out so that a pilot could accurately read and engage them during the high stress of aerial battle. Moreover, not just anyone could operate a high-performance aircraft. They had to be both intelligent and highly motivated. Physically, they must be able to withstand high G-forces while maintaining their sense

FIGURE 1.2 Cockpit of a World War II-era B-24 bomber. Note the complexity of controls and displays. (Photo courtesy of the National Museum of the US Air Force.)

of orientation. Training was lengthy and complicated. Psychologists came to play a principle role in ensuring the right soldiers found their way to the cockpit.

World War II also stimulated an exponential growth in clinical psychology. Combat stress casualties were high, and there simply were not enough psychiatrists to treat afflicted soldiers. Psychiatrists, who hold a medical degree, could not be trained quickly enough to take up the slack. This opened the door for clinical psychologists, who typically hold a doctor of philosophy degree, to expand their important role in mental health care. This role continued to grow following the war, with the establishment of the Veterans Administration (VA). The VA was charged with providing health and other assistance to the millions of men and women who served in World War II, a role that continues today.

MILITARY PSYCHOLOGY TODAY

Military Laboratories

Organizations pay for what they value. Militaries around the world continue to provide funding for psychological research. In the United States, each of the services maintains laboratories that conduct psychological research. These organizations employ civilian and uniformed military psychologists. The US Army, Air Force, and Navy all maintain their own research centers that conduct military psychology research. Collectively, these laboratories and research centers lead the way in improving how we select, train, and lead soldiers in combat, and how we deal with those who suffer from the psychological impact of war.

It is not just the United States that places great value on military psychology. I recently gave a keynote address in New Delhi at an international conference on military psychology sponsored by the Indian Ministry of Defence. Military psychologists from 14 countries and three continents attended the conference. I was struck by the common needs and interests from this group representing a truly diverse set of nations in terms of economic and political dimensions. This conference is not unique. Every year, organizations like the International Military Psychology Testing Association host meetings that bring together both uniformed and civilian military psychologists from across the globe. Now, over 20 years since the fall of the Berlin Wall, these meetings include psychologists from Russia and other countries that were once our adversaries. At West Point, we routinely host visitors from the Republic of China, and work with our counterparts from Europe, Australia, South America, and Asia on various topics that affect how we leverage psychology to improve our respective militaries. Military psychology is truly a global venture, and recognized by all world powers to be essential to the optimal functioning of both individual soldiers and the military units and organizations in which they work.

Direct Funding to Universities

The military also supports psychological research by providing funding to universities and other nongovernment research facilities. Some of the most exciting developments in psychology are funded in this manner. As we will see in Chapter 5, the US Army is funding former APA president Martin Seligman, a professor at the University of Pennsylvania, to develop a revolutionary approach to *prevent* posttraumatic stress disorder (PTSD) and other maladaptive reactions to combat stress, and to enhance a soldier's chance of experiencing posttraumatic growth (PTG) from the adversities experienced in combat.

Funding to Nongovernment Research Organizations

The military also turns to private companies to conduct much of its research. These companies may range from small businesses with only a few employees to giant corporations. To accomplish this, specific research needs are identified and published. Companies and corporations with expertise that could answer the research question can then submit contract proposals, which are awarded competitively based on the merits of the proposals. Some forms of this sort of research funding explicitly require partnerships between the military, commercial companies, and universities.

The most exciting aspect of this ongoing research is that the advances in psychological science and practice that emerge from the military are equally applicable outside the military. If we can select better soldiers, then we can select better police officers, firefighters, and insurance sales personnel. If we can design a cockpit that a fighter pilot can operate successfully under the stress of combat, we can also develop a cockpit that commercial airline pilots can employ more safely and efficiently. If we can help soldiers avoid PTSD and derive positive meaning from their trials, then we can develop ways of helping survivors of natural disasters deal more positively with their own hardships. Ironically, perhaps, research that is inspired and funded to make war more effective may result in positive impacts to everyday life outside the military, a topic we will explore in more depth in Chapter 12.

WAR HAS ALWAYS STIMULATED SCIENCE, AND VICE VERSA

If anyone is an expert on the role of psychology in war, it is retired US Army Major General Robert H. Scales. A graduate of West Point's Class of 1966—the class that suffered the heaviest casualties during the Vietnam War—Scales commanded two different units in Vietnam and was awarded the Silver Star for his actions in the battle of Hamburger Hill, later the subject of both a book and a movie. As Scales continued his military career after Vietnam, he earned a doctorate of philosophy in history from Duke University. He retired in 2000 after serving as the Commandant of the US Army War College, and since then has been outspoken in his view that psychology and related sciences are critical to success in modern war. Scales discusses the role of "amplifiers" in warfare, a concept he attributes to Ohio State University historian Alan Beyerchen.[6] In this view, every major war since the turn of the twentieth century has been associated with exponential advances in science. These amplifiers give an edge to the country that develops them or applies them most adroitly in battle. The chief amplifier in World War I was chemistry. Advances in explosives and in chemical weapons greatly increased the lethality of war. In World War II, the amplifying science was physics. The two most notable examples are the development and application of radar and the atomic bomb. These developments were of monumental importance in that war, and it is arguable that radar was the most important. Although Germany also developed radar, the United States and Britain learned how to use the technology to the greatest advantage. To put these two developments into perspective, some historians maintain that while the atomic bomb ended the war, radar actually won it.

The Cold War, which Scales and Beyerchen call World War III, represents an interesting evolution of war. With the advent of nuclear weapons and other weapons of mass destruction, the traditional war between states consisting of large armies lining up face-to-face on battlefields was no longer viable. Overt warfare during World War III was confined to regional conflicts, including Korea and Vietnam, where contrasting political ideologies were fought out in a limited manner and in no small part using proxy armies. Scales argues that World War III drove the development of information technology, including computers and digital command and control systems. The ability to rapidly detect events in the world and to process and disperse information prevented either the United States or the Soviet Union from being able to launch a surprise attack on its adversary. One need look only to Libya and Egypt today to see that modern information technologies continue to shape both society and warfare.

The Global War on Terror (GWOT) is thought of as World War IV. Scales fervently believes that the amplifiers in this war are the behavioral and social sciences. This war is not being fought among nation-states, but among competing political, religious, and social ideologies. It cannot be won with kinetic energy weapons and by simply killing more of the enemy than they kill of you. This strategy was effective in the Battle of the Bulge and in the bombing of Hiroshima. In World War II it was crystal clear who had attacked the United States. Germany attacked Poland, Russia, and Western Europe, and murdered millions of innocent civilians. Japan bombed Pearl Harbor and fought the United States across the Pacific for over three years before the atomic bomb destroyed both their ability and will to continue the war.

There is no similar easy kinetic fix to the current war. No legitimate foreign power destroyed the World Trade Center or attacked the Pentagon on September 11, 2001. We could not launch a retaliatory strategic strike (either nuclear or with conventional weapons) against an offending country. United Nations sanctions do not work against a dispersed terrorist network. There are no harbors or ports to blockade, transportation systems to disrupt, or armies to destroy. But there certainly is an enemy intent on waging war against the United States and other western countries.

Put yourself in the place of the commander in chief and his military advisors on September 12, 2001. The military toolbox contained only kinetic energy tools, and these were soon employed in both Afghanistan and Iraq. Iraq had a strong and well-trained traditional army, and it was quickly defeated with kinetic-based weapons and tactics. Once that was accomplished, the war turned into what the military now refers to as "asymmetric" warfare. This type of war is characterized by guerilla tactics, terrorism, and ideological manipulation. It is not clear who the enemy is. For the most part, military operations in Afghanistan have been this way since military operations commenced there in October of 2001.

Scales suggests that asymmetric war will be won by nations that learn to merge military firepower (there are, after all, enemies that must be killed) with a keen understanding of the psychology, sociology, and anthropology of both the enemy and the cultures in which they operate. In an era of instant worldwide communication, the operational win of killing an insurgent leader by using a remotely piloted drone to blow up his headquarters is quickly negated if that headquarters is also his residence and pictures of his dead children and wives are posted on the Internet hours after the attack. Contrast this outcome with the way the killing of Osama bin Laden was handled. He was killed without harming large numbers of noncombatants, and great pains were taken to treat his remains with proper respect and in accordance with Muslim tradition.

The role of psychology in warfare is vital. It takes a new kind of soldier to fight a new kind of war. Industrial/organizational psychologists must learn to identify the abilities and aptitudes that will allow soldiers and their leaders to know when, where, and how to employ deadly force. Engineering psychologists must design new military systems that allow the human to employ his or her strengths (sense-making, intuition, rapid decision making) and capitalize on the strengths of digital technologies (vast information management) to manage command and control systems. Modern war, like others throughout history, places tremendous psychological strain on combatants and their families. Psychologists must continue to develop innovative ways to both treat combat-stress casualties and to prevent them. Psychologists must team with other social scientists to build a true understanding of the culture and values that reside in the resident populations where soldiers are deployed. Social psychologists will tell us how to win hearts and minds, how to negotiate with enemy warlords, and how to lead armies that are comprised of what in today's terms would be nontraditional soldiers. Finally, behavioral neuroscience will combine with other sciences to perhaps develop ways of producing soldiers with almost superhuman powers.

PURPOSE OF THIS BOOK

In this book, I argue that psychology and its closely related disciplines are becoming ever more vital to the success of war. The intensity and lethality of modern weapons; the influence of near real-time digital communications systems; revolutionary developments in science, technology, and engineering; and changes in the political makeup of the world necessitate a greater understand of the human element of war than ever before. Nations and their armies that understand this fact and are able to harness the power of cutting edge psychological science will have a decided strategic advantage in wars of the twenty-first century. No longer does simple brute force or the ability to dominate the kinetic energy fight (winning based on superior firepower alone) guarantee success. One need only look at the domination of US military force over the Iraqis in March and April of 2003, where US tanks, artillery, airpower, and infantry quickly subdued the Iraqi military. In short order, "mission accomplished" turned into an interminable political and military quagmire from which the United States has yet to fully emerge. Still not convinced? Then think about how the two greatest superpowers in history—the Soviet Union and the United States—failed to win decisive military advantage in Afghanistan, despite massive technical superiority.

The following chapters focus on the role of psychology in future wars of this century. I will attempt to project what psychological science may look like in 2050, and the role it may play in wars of that era. I readily admit that predicting the future of war is risky business. I served as an air force officer during the Cold War. We equipped ourselves and trained for a confrontation between superpowers that, thankfully, did not materialize. Who could have envisioned, in 1980, an enemy that lacks a country or even a traditional military being able to bring down the World Trade Center towers and to attack the Pentagon, wreaking havoc on the United States mainland that Germany and Japan could only have fantasized about merely a generation before? Perhaps today's ideologically grounded war based on terrorism is an aberration, and sometime in this century large and technologically advanced nations will again wage war against one another. But in either case, psychology will play a pivotal role in success or failure.

Although I focus on war in the mid-twenty-first century, along the way I will explain how psychology impacts current war. Inevitably, as I move from current war to future war, I will move from a very sound empirical and scientific treatment of the role of psychology to war to a discussion that must by necessity involve some speculation. I was educated as an experimental psychologist, and as such was always admonished not to "go beyond my data," so I approach this task with some trepidation. But discussing the current state of the world is only mildly interesting, and I think that there is potentially greater value to invoking a certain degree of scientific license to explore how our rapidly growing understanding of the genetic, neurobiological, cognitive, psychological, and social bases of human behavior will interact with future technical, social, political, and economic conditions to impact wars of the mid-twenty-first century.

A COMMENT ON THE ETHICS OF SCIENCE AND WAR

The role of psychologists and other behavioral and social scientists and their involvement with the military is controversial. Some psychologists find the notion of supporting the military morally abhorrent. As we will see in Chapter 6, some anthropologists vehemently object to assisting the US Army with the fielding of human terrain teams (small teams of behavioral scientists that help a commander understand the local social and cultural conditions). I do not want to marginalize the heartfelt opinions and values of these people. It is important to have among our colleagues those who can offer another perspective. It is important for the reader to know, however, that because one is a military psychologist does not mean that he or she is free to employ psychology toward any ends dictated by the military. I have been a military psychologist for over 30 years and have never heard of or known a psychologist to be instrumental in the application of torture during the interrogation of prisoners, for example. There is a minority of psychologists who make such claims, but they do so without any knowledge of how psychologists actually are employed by and in support of the military. I hope that this book will help clarify this role, both for other psychologists and for the general reading public.

All will agree that the application of psychology or any other science in war is an enterprise fraught with potential ethical dilemmas. Physicists, who harnessed the power of the atom to produce the first and thus far only atomic weapons used in war, were and continue to be divided in their view of the use of nuclear weapons in warfare. Biochemists have developed chemical weapons that can wipe out large populations in short order. And the humanities are not exempt from these dilemmas either: Theologians and philosophers have developed religious and moral systems that encourage and promote aggression between nations and ideologies.

I will return to this topic in the final chapter. An important outcome of this book may be the stimulation of further thoughtful and informed discussion on the role of psychology specifically, and science in general, in the support of military operations. One thing is certain: War, along with opposable thumbs and language, is a species-specific behavior among *Homo sapiens*. The APA or other scientific and professional societies may someday decide to ban their members from assisting the military, but war will certainly continue. If you believe that freedom and democracy are superior to societies that repress and subjugate their own citizens, then the application of scientific technology including psychology to war may be construed as a positive contribution maintaining our way of life and the values that we as a nation hold closest to our hearts.

NOTES

1. Barbara Kingsolver, *Animal Dreams*, 14.
2. Edwin G. Boring, *A History of Experimental Psychology*, 2nd ed, ix.
3. Jonathan Shay, *Achilles in Vietnam*.
4. Sun Tzu, *The Art of War*, 18.
5. For a discussion of the role of Yerkes and psychology in World War I refer to Martin E. P. Seligman & Raymond D. Fowler, "Comprehensive Soldier Fitness and the Future of Psychology," 86.
6. Robert H. Scales, "Clausewitz and World War IV," S23–S35.

2

THE RIGHT SOLDIER FOR THE RIGHT JOB

I would say some are born. A person can be born with certain qualities of leadership: good physique, good mental capacity, curiosity, the desire to know. When you go to pick out the best pup in a litter of bird dogs, you pick out the pup even though he is only 6 weeks old. He is curious, going around looking into things, and that kind of dog usually turns out to be the best dog. But there are qualities one can improve on. A thorough knowledge of your profession is the first requirement of leadership and this certainly has to be acquired. Observing others is important—trying to determine what makes them stand out. That's why I think we can learn a lot by studying past leaders. Studying Lee, other Civil War leaders, Jackson, Lincoln. Trying to see what made them great.
General Omar Bradley[1]

When I applied to be an air force officer, I was given the Air Force Officer Qualifying Test or AFOQT (as you may know, the military *loves* acronyms!). The AFOQT measures the aptitude to serve as an air force officer. In general, aptitude tests are used to predict who will excel in a given setting or on a specific sort of task. In this way they are somewhat different from traditional intelligence tests, which purport to measure one's general cognitive ability, and generate a so-called IQ score. To the test taker, it is hard to tell the difference between an aptitude test and an intelligence test. The AFOQT generates composite scores in five areas: pilot, navigator, academic, verbal, and quantitative.

Like most people who take tests like this, I viewed the AFOQT as a hurdle I had to overcome to achieve my goal of an air force commission. The test-taking gods smiled on me that day, and I scored high enough that my recruiter said I was a VIP candidate and that he could expedite the paperwork needed to get me into Air Force Officer Training School (OTS). In retrospect, and with what I know now about recruiting, he may have invented the VIP label to boost my ego and make me more likely to accept an offer. It worked. My recruiter also said that since I was being selected to be a science officer, I would not have to run and do the other physical training one normally associates with military training. In fact, he said that I could play golf while everyone else was out doing PT. As you may have guessed, he was mistaken on this point.

Within two hours of arriving at Lackland Air Force Base for training, I found myself on a unit run with the other OTS newcomers. A recruiter who didn't tell the truth? Who knew?

In addition to the AFOQT, the air force used other criteria to select me for OTS. They inspected my undergraduate and graduate college transcripts; I had to pass a background check to ensure I did not have a criminal record; and I was given a physical to rule out any medical conditions that would prevent me from serving effectively as an officer and to ensure that I was fit enough to complete the physical training required of an officer (hum, what about that golf thing?). In terms of psychological testing, however, the AFOQT was it.

A LITTLE HISTORY

The quote by General Omar Bradley hints at the challenge faced by the military—how to select its future leaders, both enlisted and officers—and to what extent their potential is inborn or consists of a teachable set of behaviors. Since World War I, the military has turned to psychology to help it select the sorts of men and women who can serve effectively in the armed forces, as well as those who have the potential to become leaders.

The history of psychological testing to select soldiers is quite interesting, and it gave a boost to the still-developing science of psychology. Prior to World War I, all it took to be a soldier was the physical ability to successfully complete military tasks such as riding a horse, loading and firing a weapon, and completing exhausting activities such as long marches and digging fighting positions. Most of these tasks did not require the soldier to read or write or to know sophisticated math. By World War I, however, military tasks had become more demanding. Besides raw physical strength, soldiers needed to be able to operate increasingly complicated weapons systems. Motorized vehicles including tanks came into use, and it took considerable intelligence to maintain and operate them. Communications systems now included radio, in addition to telegraph. Talking on the radio may have been easy enough, but knowing how to maintain a radio required more skill. And, of course, airplanes were introduced to combat. Pilots had to quickly learn to operate and navigate the aircraft while at the same time observing and engaging the enemy.

In the United States, psychology quickly stepped up to help the military develop ways to cheaply and effectively test recruits, wean out those who were unqualified to serve, and assign those who remained to the jobs that best suited their skills. Robert Yerkes, known as the founder of military psychology, served as president of the American Psychological Association in 1917. Yerkes rallied the members of APA to support the war effort, and specifically appointed committees to, among other things: (1) develop ways of testing recruits, (2) determine how to assign recruits to jobs requiring special aptitudes such as aviation, the artillery, and communications, and (3) investigate special issues pertaining to aviation. He even formed committees to look at soldier motivation and consider how to deal with what would later be known as posttraumatic stress disorder (PTSD).

This effort resulted in the first large-scale psychological tests in history, in the military or any other context. Under Yerkes's direction and with the help of many leading psychologists of the early twentieth century, the Army Alpha test battery was developed. It consisted of eight subtests, including subtests that are found in contemporary tests such as vocabulary, mathematics, and grammar. The tests were fielded very quickly and ultimately administered to nearly 2 million men. Think of the scope of this endeavor: Not only were the tests developed and validated within just a few months, they were also reproduced in large quantities and hundreds of test administers were trained; then the tests were administered and scored at sites throughout the country; and then their results were used to classify and assign soldiers to jobs. Amazingly, this mammoth undertaking was completed without the help of computers and modern communications technology.

The Army Alpha test (and a picture-based version, called the Alpha Beta test) set the standard for future military testing efforts. In a chapter published in 2012, Army Research Institute psychologist Michael Rumsey provides a thorough and fascinating review of the subsequent history of military testing.[2] For instance, World War II spurred further advances in military personnel testing, many driven by the need to select the right people to be pilots, and these efforts continue today. But I would describe the advances in military testing since World War I as incremental improvements in the same basic approach: Identify the cognitive-based aptitudes necessary for acceptable performance, and improve the quality of the tests used to assess them. Each of the services has sponsored a tremendous amount of research into test development. Today's selection tests are far better than those developed by Yerkes and his associates, but in essence they are not much different.

These early efforts in military testing evolved into the aptitude test used today by all branches of the military, the Armed Services Vocational Aptitude Battery (ASVAB). The ASVAB is a multiple choice test that assesses the test taker's aptitude in nine areas, including verbal and mathematics knowledge and reasoning, and aptitude for skills likely to be important for those enlisting in the military such as electronics and mechanical comprehension. All applicants to the military take the ASVAB. Four areas of the ASVAB (word knowledge, paragraph comprehension, arithmetic reasoning, and mathematics knowledge) are used to compute the Armed Forces Qualifying Test (AFQT) score. The AFQT score is a percentile that classifies recruits into five general categories. For example, to be classified as a Category I recruit, the test-taker's score must place him or her in the 93rd percentile or higher. At the other end, a Category V recruit would be below the 10th percentile. US law prohibits applicants who are classified in Category V from serving in any branch of the military. Utilization of Category IV recruits (percentiles 10 through 30) is restricted and subject to additional criteria that may predict success in the military, including completion of the GED.[3]

The ASVAB/AFQT is a useful instrument, to be sure. It does a good job of initial screening in that potential recruits who score too low are likely to meet with failure in the military. This is illustrated in a unique historical experiment initiated by then-Secretary of Defense Robert McNamara in 1966. Motivated in part to both meet recruiting goals (and put less of a burden on the draft) and to make military jobs open to a broader range of men, McNamara launched Project 100,000. Project 100,000 lowered both the physical and mental standards for enlistment with the goal of bringing 100,000 additional men into the armed forces (the program was only open to men). The thought was that with proper training, these previously ineligible recruits could become effective soldiers. In general, the program was an abysmal failure. In all, over 350,000 men entered the armed forces under the auspices of Project 100,000. Although they completed basic training at nearly the same rate as recruits in higher AFQT categories, they were slower to advance in grade, ran afoul of the military justice system at higher rates, and left the service prematurely at twice the rate of other enlistees. With Project 100,000 in effect during the escalation phase of the Vietnam War, many found their way into combat jobs and became cannon fodder in the eyes of critics of the program. The military ended Project 100,000 in December 1971.[4]

There can be no doubt that nearly 100 years of aptitude testing, begun with the efforts of Robert Yerkes and his associates in World War I, has resulted in sophisticated and effective methods for the screening, selection, and assignment of recruits. However, as good as these tests are, they only assess one of many types of abilities, skills, attitudes, and knowledge that also influence how well we adapt and perform in life. Indeed, at the very most, measures of talent and attitude account for perhaps 25 percent of human performance.[5] What accounts for the other 75 percent?

WHAT IS MISSING?

Ask yourself a few questions. What makes the military a challenging and demanding job or career? What personal attributes do you think it takes to be a successful soldier, airman, sailor, or marine? If you are like most people, the first things that come to mind don't have much to do with aptitude or even IQ. Our military personnel need to be of high character, to be fair and honest in dealing with others. They must be physically and morally courageous. They must be loyal to both their country and to their units. And perhaps more than anything else, they must be tough-minded, unwilling to give up when the going gets rough. The ASVAB, AFOQT, and related tests fail to measure any of these traits and attributes.

To return for a moment to my personal experience, I found that the biggest challenges of OTS required strengths that were not at all related to aptitude. Those who failed to complete training rarely did so from lack of ability. All initial military training involves socialization to a new culture, one that embraces the team and the mission above personal interests. You are given more tasks to do in a day than one can possibly complete, and hence you need to be able to prioritize and multitask. You don't sleep as much as you are accustomed, and are often physically drained. Your trainers like to surprise you with challenges when you least expect them. Within two weeks I was tired, lonely, frustrated, and ready to quit, as were all of my classmates. We were all doing well on the academic material, but were worn down from the myriad of other challenges we faced. What differentiated those who remained versus those who failed were noncognitive factors of the sort mentioned above. Most importantly, those who remained had a "never give up" attitude. We will return to this thought shortly. In any case, my selection experience did not prepare me for what I experienced during OTS, as the reader may discern in Figure 2.1.

You may be thinking, "That may have been true years ago when you entered the air force, but it can't be true today given all that psychology has learned about testing and selection." Unfortunately, you would be wrong. To get into the army, air force, navy, or marines today is not much different than it was a half century ago. The applicant must

FIGURE 2.1 The author in his first weeks of Air Force Officer Training School in 1980. Physically and cognitively capable of success, his gloomy continence reflects the "gut check" that military training provides. (Photo courtesy of the author.)

pass an aptitude test, and be a high school graduate or have a GED (for enlistment), or be a college graduate (to be an officer). He or she must be healthy and have a modicum of physical fitness. And the applicant must not have any serious criminal convictions. Notice that applicants are not screened for psychological health.

In today's military, the most careful selection process is for admission to our military academies. I have served on the West Point admissions committee and can tell you that being selected to attend this institution is a lengthy and detailed process—much more so than my experience in the air force so many years ago. Prospective cadets are evaluated in three major domains: academic, leadership, and physical. They must meet minimum criteria for each domain. Academic potential is assessed by several factors including high school grade point average, class rank, and standardized test scores (e.g., the Scholastic Aptitude Test, or SAT). Leadership is judged by a variety of indicators, including being captain of sports teams or clubs. Physical potential includes successfully completing a test that includes several component tasks. These and other measures are grouped into a composite score, called the Whole Candidate Score, and are then rank ordered. In selecting a class, the committee generally offers admission to those with the highest Whole Candidate Scores. Although this process is certainly more involved than what I experienced or what is involved in selecting enlisted personnel, it does not capture the noncognitive traits and attributes that may be fundamentally more important to long-term success than mere aptitude.

Selecting enlisted soldiers, especially in a time of war, is not as involved as selecting candidates for admission to military academies, and indeed military academy selection has evolved almost separately from enlisted selection. The military academies are first-rate academic institutions that rival the Ivy League and other elite universities in competition for admission. They receive thousands of applications to fill about 1,200 slots for new cadets each year. Filling the ranks of enlisted personnel is harder. Some estimates suggest that only one-fourth of Americans between the age of 18 and 22 qualify for the military. They may score too low on aptitude tests, not have a high school degree or the equivalent, have criminal a record, or, increasingly, they may be physically unfit (e.g., too heavy) for the military.[6]

These constraints may help explain why the military has not fielded tests that assess a broader range of abilities, traits, and competencies, especially for enlisted personnel. Testing is expensive and time-consuming. Plus, additional tests might end up eliminating applicants from consideration, something that recruiters are loath to contemplate in an era where meeting monthly recruiting goals is an ever-present challenge. But accepting men and women who are not capable of completing military training or performing adequately even if they do complete training is also an expensive proposition. In the following pages, I will explore what I think military testing will look like in the coming decades. And take note: What works in the military will work in the civilian sector. As in the case of the Army Alpha test, I predict that innovations implemented by the military in coming years will soon be found in corporate America and our education system.

Military Testing in 2030

Aptitude

Aptitude testing is important and will continue to be refined and improved. The changes in testing aptitude for the military have their roots in research begun a over a quarter of a century ago at what was then known as the Air Force Human Resources Laboratory (AFHRL), located at Brooks Air Force Base in San Antonio, Texas. Computers were getting smaller and more powerful, and this opened up the possibility of using them to administer aptitude tests rather than paper and pencil tests that were (and still are, largely) in vogue.

Under the direction of one of the early pioneers in military psychology, Dr. Raymond Christal, the Learning Abilities Measurement Program was initiated.[7] Known as Project LAMP (remember, the military loves acronyms), scientists began to compare performance between paper and pencil tests and tests administered using computers. My first assignment after receiving my commission was at AFHRL, and I had the good fortune of being assigned to assist with Project LAMP. Christal assembled an impressive team of young scientists who had recently completed graduate study at major universities, including Stanford and the University of Illinois.

We immediately found some very interesting things. First, it quickly became apparent that computer-administered testing resulted in fewer marking errors than when tests were taken with pencils and paper. In one study, we found that 25 percent of the performance in paper and pencil testing was due to marking errors. You may have experienced these problems yourself in taking paper and pencil tests. A correct answer is one that is marked in the right place on the test itself or, more typically, a bubble scan sheet. But to mark the correct answer in the right place, you have to read the question carefully, compare possible answers to the question posed, select an answer, hold that answer in your short-term memory while you find the correct place on the answer sheet to mark the answer, and then actually put the mark (using the requisite #2 pencil, without smudging it or marking outside the lines) in the right spot (e.g., a, b, c, or d) by the correct question/answer number. While this may seem simple enough to a reader who is an experienced student, consider that the bulk of military aptitude tests are given to young men and women who have only completed high school or the equivalent, and do not have extensive experience taking standardized aptitude tests. Add in the fact that they may be highly anxious about taking the test, and you can begin to see

FIGURE 2.2 Seeking civilian employment, a veteran takes an aptitude test in 1944. (Photo courtesy of ©BETTMAN/CORBIS.)

why so many errors may occur in this manner of testing. Future testing will bear little resemblance to the scenario shown in Figure 2.2.

In contrast, the same aptitude tests administered with computers resulted in far less error. Although the cognitive processes involved in reading the question and selecting the correct response are the same, the test taker can much more easily hit a key corresponding to the selected answer, while actually viewing the question and answer options. There is no need to hold an answer in short-term memory while redirecting attention from the test itself to the answer sheet, then executing relatively more complicated motor responses to record the answer. In short, the computer-based testing method yields more valid or cleaner results than paper and pencil methods.

The advantages of computer-based testing do not stop with reducing errors. Computers enable psychologists to measure mental processes that cannot be measured with paper and pencil tests. For example, with a computer it is easy to measure the amount of time needed to read and answer a question. Speed is a component of intelligence; hence the ability to easily measure it increases the power of a test. With computers, you can decide what question to ask next based on the response to the previous question. This approach, called branching, enables the tester to more carefully explore the depth and breadth of the test taker's knowledge, skills, or aptitude.

Perhaps most importantly, computer-based testing allows us to measure basic cognitive skills that go far beyond the types of word and math questions typically found in paper and pencil tests. In Project Lamp, we began to study the relationship between basic information processing capabilities and outcome criteria, such as completion of basic training. We measured working memory capacity and speed, declarative and procedural knowledge, and various perceptual skills. By the early 1990s, we found that computer-administered tests were more predictive of subsequent performance than the ASVAB.[8]

So, one future innovation for screening, selecting, and assigning applicants to the military will involve computer-based testing. This change will not simply involve taking paper and pencil tests and putting them online. Although that will be part of testing, we can expect to see computer-administered tests that, with surgical precision, map an applicant's specific cognitive skills in all domains of human information processing, including tests of perception, attention, working memory, and decision making. Done properly, these tests will provide a baseline of brain health that could subsequently be used to assess cognitive functioning following traumatic brain injury (TBI), posttraumatic stress disorder (PTSD), or other physical or psychological traumas that soldiers may encounter. More will be said about this in Chapter 5.

Noncognitive Testing

The greatest changes in military testing in 2030 will be in expanding testing from strictly aptitude assessment to noncognitive factors. By doing so, military psychologists will greatly improve their ability to place the right person into the right job. As mentioned earlier, aptitude tests may account for about 25 percent of whatever contributes to human performance. By adding noncognitive tests, we can move that percentage higher. To the extent that screening, selection, and assignment are improved, there will be substantial cost savings through improved performance and reduced attrition.

Grit

Have you ever accomplished a goal that took years to achieve? Are you a hard worker? When you fail, do you double your efforts to succeed? Your answers to these and similar questions turn out to be strongly predictive of human achievement in a wide array of domains, ranging from success in the National Spelling Bee to completion of the cadet basic training at West Point. These and similar questions make up a survey known as *grit*. Grit is not an acronym, but refers to the quality of persistence. You may recall the movie, *True Grit,* where the main character, 14-year-old Mattie Ross, embarks on a quest to avenge the murder of her father. She displays incredible grit by pursuing the killer—with the help of rogue US Marshal Rooster Cogburn—through thick and thin.

Dr. Angela Duckworth, a psychology professor at the University of Pennsylvania, created the grit survey to measure the degree of persistence that people display in achieving difficult, long-term goals. During her time as a Marshall Scholar, Duckworth informally observed that while all of her fellow scholars were highly intelligent, they were not noticeably more intelligent than people she had known as an undergraduate at Harvard. She sensed that what differentiated the Marshall Scholars from other gifted individuals was their ceaseless drive to excel, and their unwillingness to quit in the face of obstacles. When she began her doctoral studies at the University of Pennsylvania, Duckworth decided to more systematically explore this persistence-like trait, as well as other noncognitive traits that contribute to why some people excel and flourish and others—with equal talent—do not, or at least not at the highest levels. Duckworth's exploration soon led to the invention of the grit scale.

The initial studies with grit proved fascinating. Duckworth found that grit predicted success in the National Spelling Bee, and that Ivy League students with higher grit scores received higher grades than those with lower scores.[9] Educational attainment was strongly related to grit in a large sample of adults. Adults who had not completed high school had the lowest grit scores as a group; scores increased in a near linear fashion to the group with the highest grit scores, adults who had completed a postgraduate degree. In between, as educational attainment increased to high school completion, some college, and completion of a bachelor's degree, grit scores continued to rise. The only exception was in adults who reported that they had completed an associate's degree; the mean grit scores of this group equaled the scores of adults who attained an advanced degree. While this finding may seem odd, keep in mind that community college students often work full time and have families. Anyone who can balance work, family, and school must be highly motivated to succeed.

Grit also increases with age. In fact, the relationship between grit and age is stunning. People age 65 and older test so much higher than other age groups that their distribution of grit scores do not even overlap with the other age groups. Is this because people become grittier with age, or did the people who comprise the over-65 generation have different developmental experiences that led to higher grit? More research is needed to answer this intriguing question.

The most interesting findings from Duckworth's initial grit studies deal with grit and military performance. In the summer of 2004, Duckworth and her colleagues Christopher Peterson of the University of Michigan and Nansook Park (then at Rhode Island University, now at the University of Michigan) approached me about administering the grit survey to all incoming members of the West Point Class of 2008. Collectively, we felt that completing the rigorous military training that occurs during a new cadet's first weeks at West Point would be a true test of grit. Moreover, West Point collects a wealth of information about new cadets including measures of aptitude, leadership potential, and physical fitness—all of which presumably contribute to success in training and later as officers in the US Army. This would enable us to evaluate the relative strength of a host of different factors that might relate to West Point success.

Before sharing the results, let me describe what a new cadet faces during his or her first few weeks at West Point. New cadets arrive on or about the first of July, usually accompanied by their parents. An upper-class cadet, dressed in a sharply worn uniform, welcomes the new cadets to West Point. After a short briefing, the new cadets are told they have sixty seconds to say goodbye to their parents. Knowing the significance of the journey on which they are about to embark, this is quite an emotional moment for both the new cadets and their parents. At the end of this sixty-second period, the new cadets—still wearing their civilian clothing—are literally marched off to begin several days of indoctrination into the army. The new cadets are given military haircuts: Men receive buzz cuts, and the women must have neatly groomed hair that does not fall over the eyebrows or extend below the bottom edge of the collar. New cadets are issued uniforms, assigned to barracks, and along the way they are offered "encouragement" and training by upper-class cadets, collectively referred to as the cadre. They encounter the cadet in the red sash, who insists that they follow very intricate instructions on behavior and bearing to perfection. Any failure in adhering to instructions results in immediate correction. This experience is extremely intimidating to new cadets, many of whom later report that they were so fearful that they barely remember their first few hours at West Point.

For the next six weeks the cadets are trained and socialized into the army. They learn military customs and courtesies and how to wear the uniform properly. They are awakened by 5:30 a.m. and train until midnight. Rising earlier in the morning after several hours less sleep than they are accustomed to, they engage in physical and military training all day. At the end of cadet basic training, following a period of field training where they learn marksmanship and other military skills, they complete a 12-mile march back to the main cadet area. Wearing the combat uniform and carrying their M-16 rifle along with a 35-pound backpack, this culmination in their initial training presents a significant challenge. In short, for six weeks new cadets are constantly stressed and challenged as they learn to become soldiers. Figure 2.3 depicts a test common to military field training.

The grit survey was given to slightly more than 1,200 new cadets on their second or third day after arriving at West Point. Given this early in training, it is unlikely that their grit score was affected by socialization to the army and West Point. Duckworth and her colleagues compared the impact of grit versus the Whole Candidate Score described earlier in this chapter, which is a composite of academic, leadership, and physical fitness potential.

How does grit play out in this context? Grit proved to be the only predictor of the successful completion of West Point cadet basic training. New cadets whose grit scores were in the top 15 percent of their class were 60 percent more likely to complete cadet basic training. The Whole Candidate Score did not predict completion at all.

FIGURE 2.3 Soldiers complete an obstacle during a Leader Reaction Course exercise, designed to build trust and teamwork among cadets. (Photo courtesy of Stephen Flanagan.)

Subsequent analyses showed that cadets who were very low in grit not only were less likely to complete cadet basic training; they were also more likely to be referred to mental health professionals for counseling.

West Point is more than an officer training facility. It is also a top-tier institution of higher education. Cadets who successfully complete cadet basic training then begin their first academic term. The coursework is rigorous. Now referred to as plebes, first-semester cadets enroll in five academic courses including introduction to calculus, chemistry, history, English composition, and psychology or information technology. They also take physical education and military development courses. All are required to play sports, either intramurals or intercollegiate. Each is assigned to an upper-class cadet who carefully monitors, coaches, and when necessary corrects their behavior. They sleep an average of just over five hours a night. Like college students everywhere, they cope with being away from home for the first time and must learn how to study for college-level courses.[10]

In this demanding context—one that most cadets contend is much more challenging than cadet basic training—grit continues to predict success. Although the Whole Candidate Score emerges as the strongest predictor of fall semester success, grit adds to the prediction, especially for military performance measures. In a subsequent study conducted on the West Point Class of 2009, Duckworth compared grit with a standard personality test—the Big Five Inventory—in predicting West Point success. The results of the personality test did not predict completion of cadet basic training.

Hardiness

Reflect on times in your life when you overcame many obstacles to achieve a difficult, long-term goal. The eminent personality theorist Salvatore Maddi has spent years investigating what psychological traits—beyond intelligence or aptitude—are characteristic of people who hang in, persist, and rise to the occasion of life's more difficult challenges. From this work, which has included people and workers from all walks of life, Maddi developed the concept of psychological hardiness. He found that hardy people—those who seem to survive and flourish no matter the odds—consistently score higher in measures of hardiness than those who struggle in the face of adversity and stress. In addition, he has identified three components of hardiness, each of which contributes to overall success. These are *commitment*, or the ability to stick with a task and not give up easily; *control*, the sense that what happens to you in life is dependent on the choices and actions you take; and *challenge*, or viewing difficult tasks as obstacles to be overcome, not as insurmountable roadblocks that will necessarily lead to failure. The commitment facet of hardiness is conceptually similar to grit in that both measure persistence. But grit focuses more on persistence in goal achievement, like graduating from college, whereas hardiness speaks more to persistence in adapting emotionally and cognitively to life's challenges.

Consider the fictional but representative case of "Sara," a 30-year-old single mother of two children. Sara works 40 hours a week, maintains a household, and cares for her children. She gets off from work at 5:00 p.m., goes home and makes dinner for her children, and by 7:00 p.m. is at her local college where she is enrolled in courses that will lead to a bachelor's degree in business. Her dream is to eventually open her own company, and she knows that having a business degree will give her an edge. Given her other responsibilities, she can only take six courses each year. At this rate, it will take Sara nearly seven years to complete her degree. Sara knows better than anyone how difficult this task is. She is chronically tired and must carefully choreograph every minute of every day to balance her responsibilities. To top it off, attending college adds to the financial burden of maintaining her lifestyle.

Now consider the fictional case of "Ashley." Ashley has a personal situation similar to Sara's. She married immediately after graduating from high school and rather than entering college, Ashley took a job. Within five years, she had given birth to two children and was divorced. Although Ashley is getting by, she spends her free time with friends, recovering from each demanding day of work and childcare. She was a good student in high school and would like to be a social worker some day, but does not enroll in college. She feels that her circumstances prevent her from pursuing a degree; she envisions attending class so many years after completing high school as threat, and she does not feel she can commit to a long-term goal such as completing a college education.

Consider how Sara responds to this scenario compared to Ashley. Sara believes she controls her own fate. She can figure out a plan to pay for and attend college while balancing her other responsibilities. Ashley, in contrast, feels at this point in her life that she cannot be successful because the external demands of work and family prevent her from doing so. Sara views obtaining a college degree as a challenge, something that will test her ability to overcome obstacles but at the same time demonstrate her competence in being able to do so. Viewing college as a threat, Ashley fears that she does not have the ability to make good grades and that she will look dumb or foolish if she tries. Finally, Sara has the attitude that she is going to complete her degree no matter how long it takes, whereas Ashley simply cannot see herself committing to a long-term endeavor.

In this scenario, Sara demonstrates a sense of control, strong commitment, and of viewing life as a challenge. People with these traits, Maddi and other researchers have found, are more likely to excel in business and school, and be less vulnerable to stress than people like Ashley, who is low in these characteristics.[11] Maddi has developed a simple test for hardiness, and it includes questions that get at each of the "3 Cs." Here are representative questions for each of the three components of hardiness:

- "By working hard, you can always achieve your goal" (control)
- "I generally complete tasks that I undertake" (commitment)
- "I really look forward to my work activities" (challenge)

Hardiness is a very important trait in soldiers, and I expect that future selection and classification tests will include a measure of hardiness. For example, US Army soldiers who score high on hardiness fare better on deployments than less hardy soldiers. They remain healthier and report less combat and general life stress, even in the context of life-threatening combat deployments.[12] Not only do hardy soldiers adjust better during combat, they also do better than less hardy soldiers after they return from the combat deployment. Teams comprised of hardy soldiers show better cohesion, and those high in hardiness show higher ratings of leadership in demanding training exercises.

Sal Maddi and I have found that West Point cadets who are high in hardiness are more likely to complete cadet basic training than less hardy cadets, and they also out-perform them through the first year of study.[13] These effects may extend for several years from the time of testing. Hardy cadets become hardy officers, and are rated more favorably on a number of dimensions by their commanders several years following graduation.[14]

Character Strengths

I have argued that many things besides mere intelligence are important to being a good soldier. Try this little experiment. Find a current or former officer or noncommissioned officer, and ask them which of the following soldiers they would rather have in their platoon: "Jim," who is highly intelligent but lacking in character, or "Joe," who is not as smart but is courageous, loyal, and honorable. One hundred percent of the time you will be told they prefer the soldier with the stronger character. I suspect that it is not terribly different in other professions as well. I know from my time as a law enforcement officer that intangible traits of common sense, bravery, and a sense of teamwork always trumped aptitude.

None of this is any surprise to the military. For example, US Army doctrine specifies seven core values necessary for an effective soldier: loyalty, duty, respect, selfless service, honor, integrity, and personal courage. Centuries of experience have taught the army and other military services the fundamental importance of character as the bedrock of both effective performance and the successful personal adaptation to the rigors of military life. Military training focuses on instilling and strengthening these values as much as it does in developing military-specific job skills such as rifle marksmanship.

Until recently, there was no science of character to allow the military to formally define and assess personal values and character. In 2004, however, psychologists Christopher Peterson and Martin Seligman published a book that described a theory of human character as well as a method for measuring it. They proposed that there are 24 character strengths that are valued by all people in all cultures. These 24 strengths are categorized into six "moral virtues." These six moral virtues, with their associated character strengths, are *wisdom and knowledge* (creativity, curiosity, judgment, love of learning, perspective); *courage* (bravery, persistence, integrity, vitality); *humanity* (capacity to love, kindness, social intelligence); *justice* (teamwork, fairness, leadership); *temperance* (forgiveness, modesty, prudence, self-regulation); and *transcendence* (appreciation of beauty, gratitude, hope/optimism, humor, and spirituality).[15]

Moreover, Peterson and Seligman described the initial development and validation of a questionnaire designed to measure the 24 character strengths, the Values-in-Action Inventory of Strengths, or VIA-IS. The VIA-IS consists of 240 simple statements, 10 associated with each of the 24 character strengths, and yields a profile of the test-taker's strengths. Since 2004, hundreds of thousands of people from around the world have completed the VIA-IS, and it has repeatedly been demonstrated to be a reliable and valid measure of character. You may take this (and other, related tests) at www. authentichappiness.org. The test takes about 30 minutes to complete and provides a rank-ordered list of your 24 character strengths.

In the summer of 2005, Seligman hosted the Medici II Conference at the University of Pennsylvania. He brought together leading positive psychologists from around the world, and spent eight weeks exploring how positive psychology could influence large organizations. I had the opportunity to be part of the Medici II, and worked with both Seligman and Peterson in studying ways that positive psychology might be used by the military to improve the selection, assignment, training, and development of soldiers. I had felt for some time that positive psychology, with its emphasis on positive human growth, flourishing, and excellence, was a perfect match with the military, which emphasizes similar things. The military believes in giving everyone a fair chance, promotes personal growth, and embraces a philosophy that encourages each soldier to rise to his or her highest potential.

Several questions immediately come to mind. Do military members differ from civilians in character strengths? Are character strengths, as measured by the VIA-IS, predictive of soldier performance, especially in challenging training and operational contexts? Are there certain character strengths that military leaders tend to employ in life-and-death situations?

We have found that military members from different countries are more alike in terms of their character strength profiles than they are with their own countrymen. Using the VIA-IS, I compared West Point cadets with a large sample of US citizens of the same age who had completed some college.[16] The West Point cadets were higher than the civilians in several strengths that are both conceptually and doctrinally related to military success. These included bravery, honesty, leadership, teamwork, self-regulation, and persistence. Perhaps more interesting, West Point cadets were more similar in their character strengths to cadets from the Royal Norwegian Naval Academy than they were to their own countrymen. In terms of strengths of character, there does indeed appear to be a "band of brothers." This will not surprise anyone who is knowledgeable of the military, but reinforces the notion that formally assessing strengths can help identify those who will fit it and ultimately find success in the military.

I have also looked at the 24 strengths in terms of how they affect the performance of West Point cadets.[17] In general, the strengths mentioned above predict successful completion of basic training as well as performance in the academic, leadership, and physical fitness domains. In one study, I compared cadets high in specific character strengths ("eagles"), with those low in the strengths ("turkeys"). I looked at objective indicators of academic, leadership, and physical performance. In almost every case, the eagles outperformed the turkeys. In another study, cadets who completed cadet basic training tested higher on the VIA-IS in nine character strengths—bravery, zest, fairness, honesty, persistence (similar to grit), optimism, leadership, self-regulation, and teamwork—than those who failed to complete the training.

Findings like this are not unique to West Point. My collaborators and I have found that character strengths are important in the success of Royal Norwegian Navy and Army Academy cadets on various exercises and missions.[18] The same pattern of strengths seen in West Point cadets is also important among the Norwegian cadets. In short, cadets who are brave, persistent, honest, and team-oriented do better than those who are not.

What about real soldiers assigned to combat? What strengths of character are important to succeeding when lives are on the line? If you think of your 24 strengths as being something akin to a toolbox, and that you can at least to some degree select a specific tool to deal with a specific task, what character strengths are used in combat? I gave a survey to a large sample of army captains who had just returned from combat rotations in Iraq and Afghanistan.[19] I surveyed these officers at the height of military operations in both countries, and almost all of them had firsthand experience with the worst imaginable combat conditions. I gave them a short description of each of the 24 strengths and asked them to rate how important each trait was to dealing successfully with the challenges they encountered on their most recent combat deployment.

You will note that I did not try to measure what their personal hierarchy of character strengths was; rather, I was interested in the strengths they *used* to deal effectively with combat. The results were very clear. These men (all were in the infantry, where women were excluded from serving at the time) indicated that the strengths they most often turned to during stressful situations were (in descending order) teamwork, bravery, capacity to love, persistence, and honesty. These are the character strengths that matter most when dealing with the taking of lives, the death of soldiers, and the loneliness of months away from home. Soldiers depend on each other (teamwork) to get the job done. They must expose themselves to mortal danger (bravery), and care deeply about their fellow soldiers (capacity to love). They must not give up (persistence), and their very lives depend on doing what they say they are going to do (honesty).

In contrast, the following strengths were rated least important in combat: prudence, spirituality, curiosity, creativity, and appreciation of beauty. On the face of it, the low ranking of spirituality may be surprising. Everyone has heard the old saying, "there are no atheists in a foxhole." But remember, these men were asked to rate the strengths they found most important in dealing with a specific challenging incident. If asked what strengths were important in sense-making and coming to terms with life and death, they well may have indicated spirituality as critical.

What about the bravest of the brave, those soldiers who sacrificed everything in combat? Nansook Park did a content analysis of the narratives accompanying Medal of Honor citations.[20] The Medal of Honor is rarely awarded these days—just a handful in the past decade of war—and most are awarded posthumously. What character strengths did these heroes demonstrate? Of course, 100 percent of them were brave. But the next highest character strengths were self-regulation, persistence, leadership, and teamwork. Intertwined with these traits was a theme of humility and selfless service. These Medal of Honor recipients put the welfare of others ahead of themselves, valued the lives of their fellow soldiers above their own, and simply would not give up under the most brutal conditions of combat.

Tailored Adaptive Personality Assessment System (TAPAS)

The US Army is currently field-testing a new psychological screening technique that includes several noncognitive tests, the Tailored Adaptive Personality Assessment System (TAPAS). It measures, among other traits, dominance and attention-seeking. The TAPAS is administered using a computer; preliminary results suggest that it is effective in predicting a variety of proficiency and motivational outcomes, and it is now being tried in Military Entrance Processing Stations. A strength of the TAPAS is that it includes strategies to make it less subject to faking. Preliminary results clearly show that the TAPAS improves the ability to predict success among army recruits in both training and subsequent job performance. The TAPAS represents the closest thing yet to an operationally fielded selection test that incorporates noncognitive measures of the sort described in this chapter.[21]

OTHER FUTURE DEVELOPMENTS

Along with improvements to aptitude testing and assessing noncognitive traits and characteristics, modern neuroscience will likely contribute other innovations to military testing over the next 25 years. Contemporary neuroscience is rapidly mapping the genetic basis of human vulnerability to a host of diseases, including those with a psychological basis such as PTSD. Neuroscientists are also looking at the genetic and neuroendocrine bases of resilience. The new field of epigenetics is investigating the manner in which environment affects the expression of genes. Much of this work will be relevant to the military. For example, simple genetic and endocrine tests may reveal who will be resilient in the face of combat, and who will be most vulnerable to PTSD, depression, and anxiety. These biobehavioral tests will allow military testing centers to screen applicants on the basis of their genetic and constitutional proclivities for adjusting to military life. Those who are vulnerable to the psychological impact of war may be assigned to noncombat jobs, or referred to jobs outside the military. Those with exceptional resilience potential may be steered toward Special Forces and other military occupations that require the most from their members.

SUMMARY

I began this chapter with my own experiences of joining the military. I was given aptitude tests that indicated a reasonable degree of intelligence. I met basic education and other requirements. I did not have any significant physical disabilities. My eyesight and color vision were normal. Basically, I was viewed as sufficiently intelligent and fit to serve in the air force.

Let's flash forward 25 years from now. A recent high school graduate is interested in a career in the military. She uses a portable digital device (what will smart phones look like in 25 years?) and learns about the army, air force, navy, and marines. She decides she wants to serve in the army, and digitally contacts a recruiter. In the next few days, she completes a series of aptitude tests and cognitive assessment batteries online and builds her virtual portfolio. She also completes a variety of tests assessing character traits that predict success in the army, to include grit and various other character strengths. Once these are completed, she takes a complete physical that includes brain scans, tests of neuroendocrine functioning, and genetic assays. This information is then entered into a computing system that compares her unique psychological, emotional, neurological, and physical profile to jobs the army needs to fill. Shortly thereafter, the army welcomes to its ranks a new soldier who is perfectly matched to the assigned job.

The payoffs to the individual soldier are obvious. Future testing will assign people to jobs they enjoy and in which they can perform well. This will increase job satisfaction. More new soldiers may choose to remain in the army (or other service) for a career. The payoffs to the military organization are also significant. Higher retention means lower recruiting and training costs. Soldiers who are placed into optimal jobs will work together better as teams. Soldiers will be less vulnerable to combat stress, and both traditional health care and psychological health care costs will decrease. In short, the combat effectiveness of the military will rise to previously unseen levels.

This revolution in training will be incorporated by the civilian sector as well, just as simple aptitude testing was adopted in the corporate world in the years following World War I. People will find it easier to match their own strengths with jobs that utilize those strengths. As in the military, retention and job performance will improve. As it does so, people will be more satisfied and less stressed. Job-related psychopathologies (depression, anxiety, and substance abuse) will decrease. The GNP will soar!

This is no scientific dream. The basic science needed to improve testing is being done right now. It is exciting to contemplate how these changes will improve the work world, both in and out of the military, in the years to come.

NOTES

1. US Department of the Army, *Leadership Statements and Quotes*, 3.
2. For a thorough review of the history of psychological testing and screening in the military, see Michael G. Rumsey, "Military Selection and Classification in the United States," 129–47.
3. For more information on the ASVAB test refer to the Official Site of the ASVAB website at http://official-asvab.com/understand_coun.htm.
4. Much has been written on Project 100,000. For a critical review of the project, see Janice H. Laurence et al., *Effects of Military Experience on the Post-Service Lives of Low-Aptitude Recruits.*
5. See Angela L. Duckworth et al., "Grit," 1087–101.
6. See William Christeson, Amy Dawson Taggart, and Soren Messner-Zidell, *Ready, Willing, and Unable to Serve.*
7. For a excellent review of the origin and initial findings of Project LAMP see Patrick C. Kyllonen and Raymond E. Christal, "Cognitive Modeling of Learning Abilities: A Status Report of LAMP."
8. Patrick C. Kyllonen, "Cognitive Abilities Testing," 103–25.
9. Duckworth describes the development of grit and her groundbreaking research on its relationship to human performance in Duckworth et al. (see note 5, above).
10. For a description of the West Point sleep study, see Nita Lewis Miller and Lawrence G. Shattuck, "Sleep Patterns of Young Men and Women Enrolled at the United States Military Academy," 837–41. For a more general discussion of the critical role of sleep in military operations, see Nita Lewis Miller, Panagiotis Matsangas, and Aileen Kenney, "The Role of Sleep in the Military: Implications for Training and Operational Effectiveness," 262–81.
11. For a historical review of the hardiness concept, see Salvatore R. Maddi, *Hardiness.* Chapter 8 focuses on hardiness and its relevance to the military.
12. Matthews provides a review of the role of hardiness in Army contexts in Michael D. Matthews, "Positive Psychology," 163–80.
13. See Salvatore R. Maddi et al., "The Role of Hardiness and Grit in Predicting Performance and Retention of USMA Cadets," 19–28.
14. See Paul T. Bartone, Dennis R. Kelly, and Michael D. Matthews, "Psychological Hardiness Predicts Adaptability in Military Leaders," 200–10.
15. See Christopher Peterson and Martin E. P. Seligman, *Character Strengths and Virtues* for a comprehensive review of the origin and development of this theory of human character strengths and tests designed to assess, including the Values-in-Action Inventory of Strengths, VIA-IS.
16. See Michael D. Matthews et al., "Character Strengths and Virtues of Developing Military Leaders," S57–S68.
17. Michael D. Matthews, "Where Eagles Soar."

18. Much of my work with the Norwegian military is summarized in Michael D. Matthews, "Positive Psychology," 163–80.

19. The initial results of this project were presented in Michael D. Matthews, "Character Strengths and Post-Adversity Growth in Combat Leaders."

20. Nansook Park, "Congressional Medal of Honor Recipients: A Positive Psychology Perspective."

21. For a thorough review of the development and validation of the TAPAS and related measures, see Deirdre J. Knapp and Tonia S. Heffner, *Validating Future Force Performance Measures (Army Class)*.

3

TURNING CIVILIANS INTO SOLDIERS

We find that the Romans owed the conquest of the world to no other cause than continual military training, exact observance of discipline in their camps, and unwearied cultivation of the other acts of war.
Flavius Vegetius Renatus[1]

The military does a good job of preparing and training for the last war. My great-grandfather fought for the Union in the Civil War. Training was pretty simple. Over a few short weeks, he was taught the basics of drill, military customs and courtesies, how to fire a muzzle-loading rifle, and some basic offensive and defensive soldier skills. He was trained in the skills that had been successful in previous wars, but were not, as it turned out, particularly adaptive or successful in the Civil War. Within weeks of signing up, he found himself at the Battle of Pea Ridge in Arkansas. Early in the battle, he was wounded in the foot by a musket ball fired by a confederate infantryman. Although the wound was not life-threatening, he was given a medical pension of $20 dollars a month and retired from the army. He used his pension to pay for medical school and became a country doctor. After his death, his widow lived well into the twentieth century on his Civil War pension.

Training was not much more sophisticated by the outbreak of World War I. But by then, the technology of war had begun to exceed the capabilities instilled in new soldiers by the training regimens in place at the time. Learning to march, salute, form battle lines, and so forth did not adequately prepare soldiers to operate more sophisticated weapon systems and the tactics they necessitated. Although many psychologists volunteered to assist the military in World War I, most of their efforts focused on selection, as we saw in Chapter 2. Of course, the need for better training was apparent, and psychologists evidently did offer advice on training. After the war, for instance, Lieutenant Colonel L. C. Andrews published a book on applying psychology to the training of soldiers, summarizing lessons learned from the war.[2] The methods he proposed were relatively simple by today's standards, and centered on an apprentice model wherein new officers and enlisted soldiers were coached and mentored in the learning of military skills. Military drill was still emphasized, as much for the purpose of instilling order and discipline as in maneuvering troops on the battlefield. In Figure 3.1 we see a soldier outfitted with protective gear designed to allow him to survive a gas attack.

By World War II, training had become a critical component of fielding an effective military force. Military aviation was in its fourth decade, and aircraft were larger, flew faster, and carried far more sophisticated weapons than World War I aircraft. Some, such as the B-17, B-24, and B-29, required large crews to fly and navigate the aircraft and operate offensive and defensive weapons systems. Crews had to be trained to work effectively and efficiently together, and to be prepared to do so under conditions of great stress. Radar made its appearance, along with a range of relatively sophisticated command and control systems. Tanks and other vehicles were bigger, faster, and more powerful than ever before. Gone were the days when someone like my great-grandfather could join the army and be ready to fight with only the most basic instruction in weapons, tactics, and drill.

The field of psychology rose to the occasion and psychologists quickly became involved in improving military training. Complex military tasks, like operating crew-served weapons, were analyzed and broken down into component subtasks. Then

FIGURE 3.1 Failure to learn how to don a gas mask and protective gear rapidly may result in a soldier's death. (Photo courtesy of Elzbieta Sekowska; Shutterstock.)

protocols were developed to train new soldiers in the use of these systems, evaluate the effectiveness of the training, and to prepare soldiers for deploying the systems in the heat of battle. Army research psychologist Dr. Steve Goldberg points out that World War II saw tremendous developments in training gun crews, engineer room personnel in the navy, and radio operators. Most notably, the army founded the Aviation Psychology Program. Many notable psychologists, including Neal Miller, participated in this program, which supported research into the training of pilots, gunners, radio operators, and navigators.[3]

It is worth pausing to consider the scope of the problem faced by the military during World War II. In 1940, the US military had 458,365 soldiers, sailors, and marines. Collectively, the US military was among the smallest of industrialized nations at that time. By the end of World War II, there were 12,055,884 men and women serving in all branches of the military. For comparison's sake, in September of 2011, total US forces numbered 1,468,364.[4] Most of the men and women who served in World War II were true citizen-soldiers. They were not professional soldiers, and had to be quickly trained to expertly execute their given jobs. My father (seen in Figure 3.2) was typical of his generation. The "army joined him" (as he put it, describing the experience of being drafted) in 1943. A year later, he was a combat medic and was then involved in 14 months of continuous combat in the Pacific theater. The army had to quickly turn this bookkeeper into a combat medic by teaching him first aid and life-saving skills. This same scenario played out millions of times for hundreds of different military job skills. To the extent that success in war depends on the skills of its soldiers, psychologists played a heroic role in helping the nation transform from a second-rate military

FIGURE 3.2 The author's father, a combat medic in the Pacific Theater in World War II. (Photo from collection of author.)

to a world power in the space of a few months. Perhaps nowhere else in history has psychology played such a pivotal role in military success. And, of course, the lessons learned about training complex skills across the multitude of different job domains quickly spread into the corporate world following the end of the war.

After World War II, the military services established a variety of psychology research units that continued to work on a range of military psychology issues, with a strong emphasis on training. The air force, now a separate branch of the military, established the Personnel Research Directorate in San Antonio, Texas, which later became the Air Force Human Resources Laboratory and eventually part of the Air Force Research Laboratory. The army founded what would later become the Army Research Institute, and the navy established the Navy Personnel Research and Development Center. These labs collectively employed hundreds of both uniformed and civilian psychologists, many of whom focused their efforts on military training.

Training in today's military bears only partial resemblance to that of previous eras. New soldiers are taught the fundamentals of drill, just as soldiers have been taught throughout the recorded history of warfare. They are given extensive physical fitness training, taught military courtesies and customs, and learn basic skills needed to operate common weapons systems. But here the similarities end. Following basic training, virtually all enlisted personnel and officers receive lengthy and complex technical training in their job specialty. Training is a career-long process. It does not end when the recruit finishes basic training.

I will illustrate this with the experience of a new recruit to the US Army. He or she is assigned to initial entry training (IET). The IET lasts from 9 to 14 weeks and includes indoctrination into military culture. The recruit engages in daily physical fitness training and drill. She learns to fire the M-16 rifle, and learns to whom and when to salute. Throughout IET, the idea that the soldier is subordinate to the larger unit is emphasized. The new recruit becomes physically stronger and more disciplined. He takes pride in his unit and in being part of something larger than himself. Graduation day is a time for the recruit and his or her family to celebrate. Hugs are exchanged and tears shed. But the recruit is not yet ready for an operational assignment.

Because the United States maintains a professional, all-volunteer military, its goal is to train and develop new personnel for the long term. In the largely drafted army of the Vietnam War, in contrast, there was little time for additional training. New recruits completed basic training and, unless they had a very specialized occupation (such as a medic), they were assigned to an operational unit and shipped off for combat. They completed their combat tour and their short obligation of active duty, and returned to civilian life.

Returning to our example of an army recruit, following IET graduation almost all new soldiers continue with additional training for their assigned occupation. This might be in transportation, military intelligence, military police, medical services, and many, many more jobs. Only after this training is completed will the new soldier be assigned to a unit. And they are not through yet. If he or she decides to make a career of the army, they can expect to later attend separate training courses designed to prepare them for higher ranks and greater responsibilities. In a 30-year career, it is not unusual for an enlisted soldier (and later the noncommissioned officer) to spend several years in training courses of various sorts.

The same is true for officers. Newly commissioned officers (in the US military, about one fourth come from West Point, the Air Force Academy, and the US Naval Academy, with the remainder commissioned through the ROTC or Officer Candidate Schools) immediately attend "officer basic" schools that teach them specific leadership and technical skills unique to their chosen specialty. An army infantryman, for instance, will attend the Infantry Officer's Basic Course, then airborne training (if not completed earlier during his precommissioning time), and probably Ranger school. It is not unusual for an infantryman to receive six months or more additional postcommissioning training before being placed into an operational unit. The training is longer and more extensive for highly specialized jobs such as helicopter pilot or special forces personnel.

After the basic course and subsequent training, the new officer will serve an initial tour or two in operational units, and then attend additional training courses when they are captains to prepare for company command; as majors, to prepare for senior staff duty; as lieutenant colonels, to prepare for battalion command; and as full colonels, to prepare for strategic service and possible promotion to general officer. In short, the training never stops.

WHAT DOES THE FUTURE HOLD?

The reader is right to conclude that contemporary military training is comprehensive and extensive. But times are changing. Technology and the nature of the threat facing our military is rapidly evolving and putting a strain on contemporary training methods. The military excels at training officers and enlisted soldiers in specific task-oriented skills such as flying or using complicated command and control equipment, but the future will see revolutionary changes in training driven by an evolution both in the nature of war and further innovations in technology.

Despite advances in information technology, contemporary military training is not dramatically different than it was 20 years ago. So-called platform training, where an instructor delivers lessons live in a classroom or other setting, is common. By necessity, much of the training in fundamental soldier skills such as marksmanship is hands-on. Digital technologies, such as PowerPoint, are widely used for administrative training, such as updates on security procedures or maintaining personal security while traveling. In some cases, especially for aviation and naval operations, the military has some of the most sophisticated immersive simulations in the world. But aviation and naval simulations are easier to design than simulations for ground forces, and, as we are experiencing now and will continue to experience, ground forces carry the greatest burden in war. They offer us the greatest gains and the greatest potential for failure. A training revolution is looming, and the military will soon capitalize on technologies to better prepare its members for the jobs they perform.

A truism about the military is that it is always prepared to fight the previous war. The nature of the global war on terror (GWOT), characterized by combat against ideological extremists and terrorists, caught us unprepared. For decades, the United States trained and equipped its military to engage a formal nation-state and its associated military forces. There would be battle lines, recognized rules of engagement, and clear-cut military objectives. The Russians would sweep across the Volga Gap to invade Western Europe, or perhaps the Chinese would invade Taiwan. In any case, we would pit large mechanized forces against similarly equipped opponents. The military with the most kinetic power (bombs, planes, missiles, etc.) would prevail. Our military trained endlessly to prepare for this scenario.

The events of September 11, 2001 caught the United States and the world by surprise. A small number of radical extremists managed to inflict major physical, psychological, and political/economic damage on the United States to an extent that neither Germany nor Japan could muster in World War II. The United States wanted revenge, and rightly so. But against what country? The attack came not from an enemy nation, but from an international affiliation of terrorists. No foreign army or air force was involved, only suicide bombers.

The skills and tactics needed to fight the GWOT are greatly different from those needed to fight a traditional war. War is changing on almost all dimensions. The remainder of this chapter explores how these changes will impact the training of military personnel in the decades to come. Today's military training may seem as archaic to the military of 2030 as training in World War I seems to us today.

FUTURE TRAINING

Future training must take into account the changing threats and missions likely to face the military in the mid-twenty-first century. Soldiers will need to be more adaptive, more able to change roles from peacekeeper to full-out warrior in very short order. They must become more savvy about cultural matters. You cannot win hearts and minds if you do not respect the social and cultural heritage of the population you are trying to influence. The physical and mental skills needed by war-fighters will expand to include skill sets that are incongruent with contemporary military culture and values. Military training in the future will of course include many of the same elements as today, but it will expand and change in several dimensions. What follows is a look at military training in the not-so-distant future.

High-Fidelity Battle Simulations Will Expand to Include All Types of Military Missions

In Gulf War I the coalition destroyed 259 Iraqi military aircraft, including 36 shot down in aerial combat. The coalition forces suffered the loss of 52 fixed-wing aircraft and 23 helicopters. None of these losses were a result of air-to-air combat; most were due to antiaircraft fire. The Iraqi air force was well equipped with Soviet-era MiG jets, including the very sophisticated MiG-29. Why were coalition forces so successful against such a well-equipped enemy?

There are several reasons for this dominance, including superior command and control systems and stealth technologies. But when it came to an aerial fight between Iraqi and coalition aircraft, training was the difference maker. Before an American fighter pilot engages an enemy in combat, he has flown thousands of missions in high-fidelity, realistic simulations. This allows the pilot to make faster and more accurate decisions, in a battle space where the opponents are engaged at or near supersonic speeds. By virtue of hundreds of simulated air battles, the pilot builds a library of schemas (what kind of threat is this?) and scripts (here is what I do in this situation).

Retired Army Major General Robert H. Scales, a strong proponent for improved training for the US Army, calls this "bloodless learning." Simulations enable the fighter pilot to learn from his mistakes without jeopardizing his airplane, other crew, or his own life. But many of the wars and other military engagements of the twenty-first century will not rely so much on airpower as they will ground forces. The story for training ground forces is much different than it is for aviation.

Throughout the history of war—and certainly true in today's operations in Afghanistan—new small unit ground force leaders (lieutenants and mid-level NCOs) make their biggest mistakes in their first weeks in combat. In Vietnam, for example, some experts estimated the life expectancy of a new second lieutenant to be measured in terms of minutes during their first firefight.[5] If they survived their first fight, they were much more likely to survive their combat tour. New lieutenants made many mistakes, from being overly aggressive to making it obvious to the enemy that they were officers by wearing their rank or by overt use of radios, arm and hand signals, and the like. Combat is a great teacher—if you are given the opportunity for a second chance, you will seldom make the same mistake twice.

In contrast to the great strides over the past 50 years in training aviators, infantry and other ground forces soldiers learn their trade much the same way they did in the wars of the twentieth century. This includes lots of drill, field exercises, and mentoring. The army calls it the "crawl, walk, run" model, but in truth most soldiers don't get to the run stage of expertise prior to deployment into combat.

The technology exists to change this. Ground force training of the future must by necessity be heavily simulation-based. The commanding general of the US Army Training and Doctrine Command, General David Cone, agrees.[6] In October of 2011, he announced that developing effective simulations for ground forces is a priority for the army in the years ahead. The other branches of the US military have similar priorities. With a small military and a population that is casualty averse, we cannot afford to train soldiers halfway. Figure 3.3 shows soldiers solving a problem in a tactical simulation exercise.

Let's compare how an infantry lieutenant is taught to lead a platoon in combat now versus how he will be trained in the future. Now, during the Infantry Officer Basic Course at Fort Benning, Georgia, lieutenants are given both classroom and field training in basic infantry skills and small unit leadership. Both are important and will remain part of the army's training strategy. However, the problem comes with providing the lieutenant with enough realistic force-on-force training that he truly learns enough to respond flexibly and adaptively in real battle.

At the officer's basic course, classes number about 40 lieutenants. During field training exercises, the officers play various roles in an infantry platoon, including

FIGURE 3.3 Soldiers engage in a virtual military operation during an experiment conducted by psychologists of the Army Research Laboratory's Research & Development Command. (Photo courtesy of the Army Research Laboratory.)

squad leader, platoon sergeant, and platoon leader (the role the officer will in fact assume upon graduation). For example, some field-training missions include navigating through terrain in search of an objective, and engaging enemy at some point or points along the way. The officer who is lucky enough to be selected as platoon leader that day will learn valuable lessons about leading soldiers in tactical combat situations. But such training is time-consuming and is intensive in use of resources needed to execute it (other soldiers to play the enemy, expert cadre to evaluate the mission and provide feedback, etc.). As a consequence, most lieutenants receive very little of this realistic training prior to deployment to combat. They have perhaps learned to crawl, but perhaps not even to walk, and certainly not to run. This leads to poor decision making and potentially the loss of life until the new platoon leader gains enough experience in combat. In contrast to bloodless learning, this is bloody learning.

The army does have experience in the use of fairly sophisticated simulators for ground forces, but mostly for tanks and other mechanized forces. These are used extensively at the national training centers, especially for training tank crews and support vehicles how to collectively move, communicate, and shoot in coordinated operations. Decision making is a major piece of this training. Tank commanders are given various scenarios and are required to quickly assess the situation, make accurate decisions, and take appropriate actions. This is not only less expensive than conducting live training with tanks—something that takes lots of equipment, time, fuel, and personnel—but also allows trainers to challenge commanders and their crews with repeated and varied tactical scenarios.

It is no accident that US mechanized forces are extremely lethal. Their success depends not only on world-class equipment, but also on the ability to superbly train crews and commanders to face any contingency, similar to the training fighter pilots receive. Modern simulations are very realistic, not like the computer games that many of us grew up playing a generation or so ago. Not only are the visual and auditory components of the simulations very good, but developers are working to make the simulation feel and smell like battle as well.

The problem has been in developing simulations of similar fidelity for ground soldiers. In an aerial fight, there may be as few as two players (a friendly and an enemy aircraft). In tank simulations, the simulated engagements typically involve relatively small numbers of tanks—perhaps a few dozen at the most. But for infantry simulations, there are many more individuals that must be realistically accounted for in the simulation. A typical infantry platoon, for example, consists of about 30 soldiers. It is commanded by a platoon leader—usually a second lieutenant—who is assisted by a platoon sergeant. There are three squads, composed of a squad leader plus eight soldiers. And squads are further broken down into two fire teams. Into this rather complicated organizational structure you must add the enemy threat. This could easily double the number of players in the simulation, and in order for the simulation to be of value, every individual must behave in ways that mimic what may happen in a tactical engagement. Imagine the complexity of simulating higher-echelon operations. An infantry company has about 150 soldiers. A battalion consists of at least three companies plus headquarters personnel. Brigades and divisions are even larger.

To complicate things further, the infantry combat environment is very fluid. The enemy seldom does what you expect them to do, and things can change on a moment's notice. This poses a significant programming challenge in creating dynamic and therefore useful simulations. And the enemy must be intelligent, in the sense of reacting in a realistic and appropriate manner in response to moves made by the friendly forces.

The US Army Research Institute at Fort Benning developed an early "dismounted" small-unit infantry simulation. ("Dismounted" refers to a unit that moves on foot, rather than being mounted in a vehicle.) This simulator consisted of a number of individual caves that a soldier could occupy and from which maneuver, communicate, and shoot at enemy forces. (A cave is shorthand for an enclosed area, about 15 by 15 feet, with a large screen at the front on which the battle simulation is projected.) In this simulator, a soldier is designated as a small unit leader (for example, a squad leader), and directs his virtual self and his virtual squad through an engagement.

I was part of the team conducting initial research and development with this system, and the results were very promising. Soldiers were able to briefly run through several scenarios and receive real-time feedback on their performance. We found that soldiers who first trained in the simulator and then conducted similar missions in a live setting outperformed those who did not have the virtual training to start.[7]

However, the system required too much technical support to serve as a practical simulator on a widespread basis. To conduct the simulations, several technicians were needed to keep the rather complicated software and hardware up and running. Other personnel were needed to play the roles of commander, enemy soldiers (i.e., someone who operated the virtual enemy, allowing them to fire at the friendly forces, take evasive action, etc.), and to provide feedback to the participants at the conclusion of the simulation. In short, this was a great concept demonstration, but not something that could be fielded.

The Future of Simulations

In the mid-twenty-first century, the constraints on infantry simulations will be overcome. General Scales has argued for several years that simulation technology has improved enough to develop high-resolution, realistic infantry battle simulations.[8] Advanced technology is not enough, though. To be effective, these simulations must be based on sound cognitive theory that incorporates the mental processes that must be honed to respond effectively in combat, all within the context of realistic operational settings. Some of the cognitive components of effective simulations include:

- A diverse array of realistic tactical scripts
- A model of how "battle savvy" improves over time
- The ability to train intuitive thinking
- Modeling complex decision making in a high-stakes setting

Moreover, a practical and useful infantry simulation system must have the following characteristics to be widely used:

- Relatively small and portable
- Stand-alone, no need for technicians or other external support
- Durable—it will be used in all sorts of conditions by many different users
- Easy to program different tactical scenarios
- Provide real-time feedback to the trainee

General Scales maintains that both the technology and the understanding of tactical cognition now exist to allow the development of these trainers. Years of warfare have provided military trainers with the knowledge and expertise to understand the components of tactical decision making. Soldiers will always train in the field, but in the middle part of the twenty-first century simulations will provide infantry leaders and their soldiers the opportunity to be like today's fighter pilots. They, too, will have led soldiers and fought in a thousand engagements before ever having contact with the enemy.

Imagine a West Point graduate in the year 2030. Upon graduation and entering the infantry basic course, he will receive much of the same classroom-based and field training that new lieutenants receive today. But now the basic course will have numerous infantry simulators in which the lieutenant can immerse himself, and respond to literally hundreds of different tactical scenarios. These simulators will be networked together, and he will be able to "play" other lieutenants in sophisticated operations ranging from full combat replete with deadly exchanges of firepower, to negotiating with local tribal chiefs. The system will be branching so that it matches the trainee's level of expertise with the difficulty of the tactical scenario that is given. As the trainee improves in both quality and speed of decision making, he will be presented ever more demanding combat situations. Feedback will be instant and complete. Moreover, it will be automatically provided through the simulation itself, and not be dependent on the presence of a live expert observer watching every simulation. The fidelity of the simulation will be lifelike. It will look, sound, feel, and smell like a real engagement.

An oft-proven reality in war is that the army that makes the fastest and best decisions has a distinct advantage over one that is slow to react. General Scales calls this *cognitive dominance*. Because they will be extremely well prepared prior to engaging the enemy, future infantry soldiers will die at lower rates than current small unit leaders. Better decisions will lead to less fratricide or other incidents that make the headlines back home. And the skills they learn, bloodlessly, will go beyond employing deadly fire to include the development of the social and cultural skills that are so vital in today's wars, and will continue to be so in wars of the future.

Training the Warrior Heart

Killing is not as simple as pulling a trigger. The psychological cost of taking another person's life—even that of an enemy soldier—is monumental. This is especially true for infantry soldiers, who witness lethal violence at intimately close distances. People who have not served in the military may have difficulty imagining the horror and soul-shaking impact of this. I refer the reader to the book *War and Redemption: Treatment and Recovery in Combat-Related Posttraumatic Stress Disorder*. Written by Dr. Larry Dewey, a psychiatrist with the US Department of Veterans Affairs, this book describes the impact of intimate warfare on soldiers. I particularly recommend Chapter 1, "The Burden of Killing," to provide a vivid understanding of what infantry soldiers experience in combat.[9]

Tom Brokaw's "greatest generation" shared the common experience of fighting in World War II.[10] When I was a boy, not only had my father served in combat in the war, but all of my friends' fathers had also served (and certainly some of their mothers, although World War II was still largely a man's war). Not all experienced combat, but in one way or another they all endured months and years of adversity. As a rule, they were pretty tight-lipped about their experiences. My father was no exception. He described his war experience as an extended camping trip and never mentioned his own combat experiences. We did not learn about his role in the war until after his death, so he took his stories to the grave with him. Who knows what demons haunted him following his return to civilian life in 1946?

The military does a good job of teaching its soldiers to kill. But it does not do a good job of teaching them to cope with it. During combat, the esprit de corps and cohesion of the combat unit may allay the soldier's psychological reactions to killing and death to some extent, but when they return to civilian life there is usually nobody with whom they share their experiences to confide in. This is especially true today, when so few citizens have ever served in the military, let alone served in combat. A veteran's struggles can be immense. Stress-related disorders among Afghanistan and Iraq veterans are common, with as many as 20 percent suffering from PTSD or related pathologies.[11] The suicide rate among soldiers has doubled in the past 10 years. The military and the US Department of Veterans Affairs (informally referred to as the VA) struggle to provide sufficient support. As we saw following Vietnam, we may end a war, but the impact on the veteran lasts a lifetime. The HBO documentary on posttraumatic stress disorder, "Wartorn," points out that fully half the residents of mental hospitals in the years following the end of the Civil War were veterans.[12]

The personal impact of killing was made vivid to me on one occasion when I was a deputy sheriff in Greene County, Missouri. Late one night, while patrolling a park in a remote part of the county, I observed a vehicle parked in an overlook. This location was noted for drug use and other crimes, and we routinely checked people in the park, especially at night. As I stepped out of my patrol car and approached the vehicle I observed a lone male subject sitting in the driver's seat. He was holding a large caliber revolver in his right hand. I could also see that he was crying, and could smell the odor of intoxicants emanating from the car. County deputies work alone and back-up officers are usually several minutes away, so I approached the subject carefully and ordered him to put his handgun out the window. Thankfully, for both of our sakes, he complied. I secured the weapon and made sure he did not have any others, and began talking with him. I learned that he was a veteran who had served in Gulf War I. He had been a tanker and had been involved in heavy combat. In his hand he held photos of dead Iraqi soldiers whom he had killed in battle, some of them horribly disfigured and burned. This man told me he could no longer live with the guilt he felt in taking these lives, and his intent that evening was to commit suicide. He lived a long way from the VA hospital and said he didn't know where to turn for help. His family didn't understand his situation and he felt the only option was to end his life. Fortunately I was able to get him admitted to a psychiatric facility that evening.

This story captures the problems that face many of our current returning veterans. While there are services to assist those who suffer from depression and other disorders linked to their service, the approach is reactive. There is little focus on prevention, only on treating pathology once it occurs. Over 2 million men and women have served in Iraq or Afghanistan, and a good number of them suffer invisible wounds that are difficult to treat, even when services are available. We must do better, and this is one area where future training will play a key role.

A glimpse into what resilience training of the future will look like can be found today. General George Casey, then chief of staff of the US Army, launched a program called Comprehensive Soldier Fitness (CSF) in 2008. This program is based on current state-of-the-art knowledge of the psychology of resilience. I was privileged to consult with General Casey as he and his senior staff developed CSF, and he flatly stated that he expected CSF to be a fixture of the army for decades to come. I will say more about CSF in Chapter 5, but in the future we can expect that all soldiers will receive extensive training in personal psychological wellness skills and that training will be an integral part of the military's career long development model.

The CSF program does not explicitly train soldiers on how to cope with the taking of lives. Future training will do so. The intent is not to develop a cold-blooded killer, but rather to teach the soldier that killing in the context of the exercise of legitimate military power in defense of his or her country is acceptable. The reader may find this to be a harsh reality, but as long as the elected senior leadership of this country sends its soldiers to war, we owe it to them to give them the skills to make meaning of their experiences, however trying. It is a far better alternative than doing nothing and letting the former soldier suffer the consequences of having done their duty to their nation.

It is worth noting before moving on to other topics that this training will reap huge benefits not just in improving the quality of soldiers' lives, but also in terms of combat readiness of our military. There are approximately 550,000 soldiers in today's active duty army (not to mention many thousands more of National Guard and reservists). Focusing only on the active duty soldiers, there is evidence that up to 20 percent may be currently suffering from psychological trauma, as we discussed earlier. People who are suffering from such pathology are not usually able to work at peak efficiency and effectiveness, and many cannot work at all. Imagine if all these soldiers had received effective training in psychological resilience skills when they entered the army and continued to get regular in-service training while on active duty. If this training reduced the pathology rate to 10 percent, then an additional 27,500 soldiers would be fully combat ready. This number represents at least two divisions of soldiers, not a trivial number in a relatively small army with worldwide responsibilities.

I haven't even talked about the cost savings associated with this training. Soldiers who are able to deal realistically and positively with adversity will become veterans who place their experiences into context. Instead of becoming a burden to the health care system, they become productive workers and offer a lifelong contribution to the economy. Not being an economist, I can't assign a dollar amount to this return on investment, but it must be in the billions of dollars.

Cultural Skills are Critical

In future war, success or failure may be determined as much by understanding and respecting the unique needs of the people residing in areas of conflict, as through pure combat power. We have all heard the phrase "winning hearts and minds." This phrase has become politicized and is somewhat trite, but it begins to capture the problem at hand. Winning the sociopolitical war has always been important, but its role is magnified in conflicts of the twenty-first century, where there are not clearly defined enemies or nations to defeat. In contrast to the wars of the early and middle twentieth century, where opposing armies stood toe to toe and fought to the death with all manner of weapons, current and future wars will require more subtle approaches to assure victory.

Another factor magnifies the importance of understanding culture in war. With near-instant and global dissemination of videos and pictures of the war, world opinion for or against a war can shift very quickly. The imprecision of violence that characterizes war is now instantly available to anyone with a computer, smart phone, or other high-tech digital device. Before I began writing this morning, I watched national news coverage of an incident involving US Marines who allegedly posted a video on the Internet showing them urinating on the corpses of dead Taliban fighters in Afghanistan, while cracking jokes about the dead fighters. There is no doubt that such atrocities have occurred since the beginning of war, but in World War II the images did not appear on your smart phone or computer, and there were no 24-hour news services repeatedly showing the incidents. Actions these marines took are inexcusable not only for showing a blatant disrespect of the enemy dead, but also from the strategic cost in terms of angering millions of people around the world, and perhaps inspiring them to further violence against US forces or citizens. There have been too many such incidents in the current wars, most notably the Abu Ghraib atrocities. To call these actions "culturally insensitive" does not do justice to the moral, ethical, and political harm they do to our nation's efforts in these wars. These actions represent *losing* hearts and minds, not winning them.

Much less dramatic, but perhaps more important, are the more subtle effects of cultural ignorance of soldiers. Certain hand gestures or signs that we may find amusing or neutral may be highly offensive in some cultures. Knowing how to treat the noncombatant population with respect to include greeting them, understanding how women and men are treated, and etiquette in proposing and conducting negotiations is vital. Respecting tribal chiefs and honoring their role in society is necessary in gaining cooperation of the population.

Knowing these rules can be used to the soldier's advantage. A friend of mine was a commander in Iraq and would often sit in conference with local chiefs and leaders. He quickly learned that to openly carry a weapon to such a meeting was a major social blunder. To gain the respect and confidence of the people he was meeting with, he set his weapon aside before entering the premises. He also learned not to rush the process. The idea of "three cups of tea" is an important part of the culture in Iraq as elsewhere in the Middle East.[13] But if he had a stubborn or hostile person with whom to negotiate, my friend learned to make a big display of weaponry. Doing so was a blatant display of power and sometimes would convince his opponent to be more open. After all, culture is a two-way street.

The military recognizes the importance of cultural education and training among its members, but current efforts are very preliminary and relatively unsophisticated.

The military academies now require all nonengineering majors to complete two full years of language training. This may be supported by a semester abroad, where they live with a family in the host nation and learn not only to speak the language but to understand the culture. The belief is that learning to understand and respect one culture should have a general effect of inspiring an understanding of the importance of learning the culture of any country in which the future officer may someday find him or herself deployed.

The military is also funding basic and applied research into new ways of teaching cultural awareness to its soldiers. Some fairly sophisticated prototypes that capitalize on immersive virtual simulations have been developed.[14] Lessons have been incorporated into various military schools. This training and education is not completely new to the military. Special forces personnel have long been trained in the culture and language of potential adversary nations. So the military is building on this expertise and developing ways to deliver it to all soldiers.

Cultural training will become a fundamental component of all enlisted and officer training in the military of the future. Knowing the customs and courtesies (and, of course, the prohibited actions) of other cultures will be as valued a skill as rifle marksmanship or knowing how to direct artillery fire with surgical precision. The training will involve many methods, from platform (live lectures and discussions), to role-playing, to sophisticated immersive virtual simulations.

Soldiers equipped with these skills will be more effective in winning the hearts and minds. Leaders who value and appreciate these skills will inculcate and reward them in their followers. In short, soldiers of the mid-twenty-first century will be true cultural warriors—able to negotiate and even entertain at one moment, and employ lethal weapons the next. A nation that does not equip its military with these skills will find itself at a strategic disadvantage.

Physical Fitness

Physical fitness (PT) training in the military serves several purposes. It builds the strength and stamina needed to perform many military tasks, both in peacetime and war. It also builds confidence. A strong soldier knows that he or she has the physical ability and skill needed to overcome daunting challenges. And, perhaps most importantly, PT builds teams. Soldiers who endure long training marches or runs and engage in strength training together form strong bonds.

I recall the PT training I experienced in my own officer training. At the time, the air force required its members to pass a 1.5 mile run check each year. On the face of it, this is a pretty minimal standard, and the times required to pass were not very stringent. Although I had played varsity basketball in high school and continued to be physically active through college and graduate school, I had never run (without a ball being involved) and I honestly viewed passing the run check as the major barrier to my eventual commissioning. However, my more experienced fellow officer trainees coached and encouraged me, and in doing so modeled leadership and team cohesion, and I passed the run with 45 seconds to spare. I was so delighted to have passed it that I earned demerits afterward for jabbering away to a fellow officer trainee (we were forbidden to talk while in formation). I could not have cared less, because I knew then that any remaining obstacles to graduation and commissioning could be easily overcome.

My other experiences during Officer Training School illustrate the importance of PT in building confidence and team cohesion as well. We were required to complete a fairly daunting obstacle course. At first glance some of the obstacles seemed impossible. I recall one that consisted of a wood beam structure about 25 feet high, set at such an angle that ascending required clinging to the beams to avoid falling backward to the ground. Nobody in their right mind would try to climb this thing. But there it stood, and up and over I went, again with encouragement and instruction from my teammates. I felt 10 feet tall at the end of that course. And the experience brought my team even closer.

The modern military faces some tough issues with respect to PT, however. The population of high school graduates, from which almost all enlisted personnel are drawn, are increasingly out of shape and overweight. Common PT tasks such as push-ups, sit-ups, and running that posed little difficulty to previous generations are major challenges to many of today's youths. The army has even waived passing all components of basic training PT tests as a requirement for completing initial entry training, in hopes that with additional time in the technical schools following basic training new soldiers will have time to get sufficiently fit and lose weight to the point they can pass the PT test. It is quite a dilemma for military trainers. Physical conditioning is an important component of military life, but if you push recruits too hard they will injure themselves and either not complete training or be turned back to another training cycle. Either course of action is expensive.

Compounding the problem is that many of the essential skills that the military needs now and will increasingly need in the future do not require a high degree of fitness. Flying an unmanned aerial vehicle (UAV) requires good eye-hand coordination and a host of perceptual and cognitive skills, but there is no relationship between physical fitness and the ability to execute this task well. You could argue that gamers, or teens and young adults who spend a lot of time on the couch playing games, might make the best UAV pilots. With the increasing importance of these and other digital technologies and tasks in all phases of the military, the relevance of a one-size-fits-all PT standard may be called into question.

So in the future there will be radical changes in PT training. Because of the valuable, although indirect, importance of the psychological benefits of PT in terms of team building and instilling self-confidence, initial military training will continue to emphasize running and calisthenics, although the standards may be altered to reflect the fitness level of incoming recruits (after all, what is important in terms of the psychological benefits of PT is the challenge, not some specific standard per se).

Once soldiers get past initial training, there will continue to be PT standards to assure some baseline of both physical competence and health, but given the increased diversity of job skills needed to maintain a technologically competent military, in the future we can expect PT standards to be adjusted to specific military occupational specialties. Infantry soldiers need more strength and flexibility than UAV operators. Pilots of traditional aircraft have different requirements than intelligence officers. These changes reflect a radical change in military philosophy, but they will be necessary to build and maintain the skill sets needed in the future military.

CONCLUSIONS AND SUMMARY

Training is the backbone of the military. In peacetime, training creates a force that is ready and able to project military power anywhere in the world in short order. In times of war, training provides a pipeline of new soldiers ready to enter the fray, and allows experienced soldiers to pass along their knowledge and lessons learned to new personnel.

Future training must by necessity change to reflect the realities of modern warfare. Building competence in all domains of soldier skills will be essential to maintaining an effective military force. It is no longer enough that soldiers be physically strong and competent with basic soldier skills like rifle marksmanship or throwing grenades. They must be culturally aware, possess the psychological body armor needed to withstand the adversity and stress of warfare, and have physical skills that are matched to the specific demands of their occupational specialty. Military trainers must learn to leverage the most modern technologies to create realistic training systems. Future soldiers will not only train in the heat and mud of outdoor ranges, but also in state-of-the-art virtual simulators that will pervade the military.

Although education is something different from training, it is worth noting that the future military will show a renewed emphasis on intellectual growth among its leaders. Napoleon Bonaparte said, "There are but two powers in the world, the sword and the mind. In the long run, the sword is always beaten by the mind."[15] A nation that invests in the intellectual capital of war will ultimately achieve much more than one that relies primarily on military might. One can easily craft the argument that avoiding or preventing war through a thorough understanding of potential adversaries is or should be the chief objective of the military of the future. Although such an approach is more difficult and time-consuming, in the end it is much better than solving every international crisis with infantry, artillery, and naval and air power. However, until that time comes, we may return once again to the lessons learned from the Roman Empire cited at the beginning of this chapter. Given the complexity of current and future military systems, the challenge of training and maintaining a highly skilled and well-led military is far more daunting a task than faced by the Romans.

NOTES

1. Flavius Vegetius Renatus, *The Military Institutions of the Romans*, www.searchquotes. com/quotation/We_find_that_the_Romans_owed_the_conquest_of_the_world_to_no_ other_cause_than_continual_military_tra/173718/.
2. Lincoln C. Andrews, *Military Manpower*.
3. For a thorough review of early military training see Stephen L. Goldberg, "Psychology's Contribution to Military Training," 241–61.
4. See the "Active Duty Military Personnel, 1940–2011" chart in the Information Please Database for a report and analysis of US military strength since the beginning of World War II, accessed December 17, 2012, www.infoplease.com/ipa/A0004598.html.
5. The question of the life expectancy of a platoon leader, usually a second lieutenant, in Vietnam is probably subject to some degree of historical distortion and reliable data are hard to find. However, survivors concur that for those encountering their first firefight, inexperienced platoon leaders were at great risk. For instance, see http://wiki.answers. com/Q/What_is_the_life_expecantcy_of_a_2nd_Lieutenant_in_Vietnam, which estimates the life expectancy of being about 16 minutes.
6. For example, see General Cone's comments as summarized in Lance M. Bacon, "General Presents His Vision for Future Warriors," *Army Times*, October 2, 2011,www.armytimes. com/news/2011/10/army-general-presents-his-vision-for-future-warriors-100211/.
7. For instance, see Robert J. Pleban et al., *Training and Assessment of Decision-Making Skills in Virtual Environments*.
8. See the Armed Forces Journal article "Small Unit Dominance: The Strategic Importance of Tactical Reform," for a more in-depth account of Major General Scales's views on the future of infantry combat simulations, at www.armedforcesjournal.com/2010/10/4757970/.
9. Larry Dewey, *War and Redemption*.
10. Tom Brokaw, *The Greatest Generation*.
11. Estimates of PTSD rates associated with Iraq and Afghanistan vary considerably depending on how the disorder is defined and a host of other factors, including the type of unit the soldiers serve in. For instance, elite and highly trained units appear to show substantially lower rates of PTSD than traditional combat units. In modern war, everyone in the theater is at mortal risk, so even jobs traditionally classified as noncombat, like transportation, put soldiers in harm's way. For a general discussion and estimate of PTSD rates, see www.rand.org/news/press/2008/04/17.html, but I encourage the interested reader to do your own research and you will find quite a disparity among estimates.
12. Jon Alpert, Ellen Goosenberg Kent, and Matthew O'Neill, *Wartorn 1861–2010*.
13. See Greg Mortenson and David Oliver Relin, *Three Cups of Tea*. The contention is that in Arab culture, there are certain social customs that should be practiced before formal

discussions commence. Sharing three cups of tea allows each side to get to know each other, builds trust, and establishes a sense of respect between the negotiating parties.

14. For example, the following report summarizes a promising approach to using simulations to train cultural skills in military personnel: Randall W. Hill Jr. et al., *Pedagogically Structured Game-Based Training*. www.dtic.mil/dtic/tr/fulltext/u2/a461575.pdf.

15. Quora Quotables, accessed December 17, 2012, www.quora.com/Quotables/There-are-but-two-powers-in-the-world-the-sword-and-the-mind-In-the-long-run-the-sword-is-always-beaten-by-the-min.

4

COGNITIVE DOMINANCE: SOLDIERS AND SYSTEMS THAT OUTTHINK THE ENEMY

No good decision was ever made in a swivel chair.
General George Patton[1]

The ability to make fast and accurate decisions is fundamental to success in war. This was true in ancient warfare and will continue to be true in the future. With the speed and lethality of modern weapons systems, the role of cognitive science, in identifying the cognitive processes necessary for military decision making, is all the more critical. Contemporary and future command and control systems allow for a near real-time picture of the battlefield, and set the stage for quick and accurate attacks on dynamically changing targets. The capabilities of both weapons and command and control systems often exceed the ability of military commanders to render the decisions needed to act and to relay that information into the hands of those who execute action against targets.

The importance of quick decision making is clear when you compare the pace of battle in wars of the twentieth century to those of today. It took relatively long periods of time for scouts and intelligence assets to determine things like enemy location and intent, and to communicate that information back to commanders who were tasked with acting on that information. In turn, it took commanders relatively long periods of time to issue orders, get them into the hands of the supporting units, and execute an attack. By the time an attack occurred, the situation may well have changed. Drastic and sometimes undesirable outcomes often followed.

Flash forward to the current war in Afghanistan. Satellites, spy planes, and drone aircraft relay live pictures to battlefield commanders. Equipped with the latest digital command and control technologies, the commanders can respond almost instantly to specific threats. The use of unmanned aerial vehicles (UAVs) to kill high-priority targets is a good example. Hesitation compromises the ability to respond to the target, which may be lost to view or may enter a civilian area where the risk of killing noncombatants is too high. Imagine the impact on the outcome of World War II if the

United States had access to the information they have now. The war probably would not have lasted more than a few months.

The need for a systematic understanding of military decision making precedes the application of formal psychological science to its understanding. By the time of the Korean War, for example, military aircraft were powered by jet engines and their speed of engagement meant that any hesitation to act—even in seconds and sometimes fractions of seconds—might result in defeat in the engagement. This need was recognized by Colonel John Boyd, a US Air Force pilot. Colonel Boyd developed a fairly sophisticated—for its time—model of decision making. He called it the OODA loop, using the acronym for the components of military decision making: observe, orient, decide, and act. Boyd suggested that pilots must be trained to complete the OODA loop elements quickly and by doing so they would be able to, in essence, outthink the enemy and give themselves a decided tactical advantage in aerial warfare.[2]

A closer look at the OODA loop model reveals interesting elaborations on each of the four components that give a hint of current and future models of military decision making. Figure 4.1 shows a detailed illustration of the model. In particular, take note of the "orient" component. This takes into account psychological variables of genetic heritage, cultural traditions, previous experience, analyses and synthesis (i.e., problem solving ability), and new information (what today we would call working memory). These processes interact with the "observe" component to yield a decision to act followed by the act itself. Note the extensive feedback loops among the four components. As one would expect, the OODA loop is ever-unfolding. Actions cause reactions by the enemy, which require adjustments by the pilot, and so on. The loop does not end until one pilot shoots down the other or the engagement is broken off for other reasons.

An interesting aspect of the OODA loop model is that it is predicated on the idea that decisions are rational and well informed. To the extent that the operator can see all of the pertinent information and builds a complete and accurate picture of the battle, then

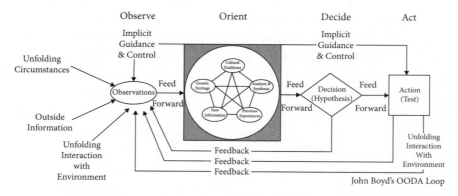

FIGURE 4.1 The OODA Loop Model of Decision Making. (Image courtesy of Patrick Edwin Moran.)

this may be a valid assertion. In fact, all current models of military decision making are based on this rational decision-making model. As we will see, this model does not apply to all circumstances in which soldiers operate, and future training in decision making will recognize that intuition and sometimes less rational decision making occurs in war.

CONTEMPORARY MILITARY DECISION MAKING

The formal decision-making model employed by the US Army is called the Military Decision Making Process (MDMP). The other services have similar models. Like the OODA loop, it was developed by practitioners of war and not by cognitive scientists. Also like the OODA loop, it assumes a rationally based decision-making process. Army Field Manual 101-5 explicitly states that the MDMP allows commanders to "reach logical decisions."[3] It can be applied at different echelons, but doctrine suggests it is most appropriately employed at levels high enough to enable the commander to assign his battle staff to develop the plan. The field manual claims several advantages to the MDMP, including the analysis and comparison of different solutions to the problem (known as Course of Actions, or COAs), well-integrated missions that synchronize fire power on the target, and the generation of a detailed operation plan. The field manual acknowledges that the chief disadvantage of the MDMP is that it is very time-consuming.

It is important to emphasize that the MDMP is grounded on the assumption that commanders have good information about the enemy, fully understand their situation, have time to prepare COAs and to critically analyze and compare their relative strengths, and have time to execute a battle plan based on this analysis. It is worth noting that higher echelons, like brigades and divisions, have more time to plan and prepare than lower echelons that are involved in dynamic tactical engagements. I have pointed out before that history prepares the military to fight the last war, not future wars. The MDMP, as we will see, is well suited to traditional warfare between nation-states where the battlefield contains actual front lines, soldiers wear uniforms, and there are clear cut objectives.

The MDMP consists of seven steps:

Step 1: Receipt of mission
Step 2: Mission analysis
Step 3: COA development
Step 4: COA analysis
Step 5: COA comparison
Step 6: COA approval
Step 7: Orders production

Each step requires considerable time and attention to detail and is dependent on the accurate completion of the previous steps. For example, mission analysis alone requires 17 tasks, including analysis of intelligence information, risk assessments, and review of assets and possible constraints. The other steps have similar subcomponents. By following the specified tasks, commanders can develop a thorough and complete battle plan.

There are several threats to the successful completion of the MDMP process. The enemy may change their strength or intent to fight, the weather may impact the assets a commander can employ, friendly units may be overtaxed and not at full combat strength, or the plan may be based on faulty intelligence information. The MDMP recognizes that with shorter planning times or for missions that are more dynamic in nature, the MDMP process may prove more challenging. Interestingly, however, doctrine as described in Field Manual 105-1 says that the MDMP must be employed even under these conditions. When time is short, the commander may assume a greater role in generating and evaluating COAs, develop fewer COAs, and lessen the level of detail and coordination in the orders that are generated. The manual says the MDMP is to be employed—in all of its seven steps—even in dynamic, rapidly changing situations typical of small unit combat. "There is still only one process, however, and omitting steps of the MDMP is not the solution," the field manual admonishes.

As good as the MDMP is for generating systematic plans against known threats, it fails to recognize how decisions are actually made in high-threat situations, and is not relevant to the full range of missions that soldiers face now and will continue to face in the future. There is no well-defined enemy in Afghanistan and Iraq. Taliban fighters shoot at our soldiers in Afghanistan and then fade back into their community, masquerading as seemingly peaceful farmers or merchants with whom our units interact. Planning and leading an armed assault on an enemy camp is a different task than helping a community plan and build a sewage disposal system while at the same time providing protection against attacks by insurgents.

The applicability of a rational model of decision making to these situations is also questionable. Cognitive scientists now recognize the use of heuristics and intuition in making decisions, especially in situations where facts are missing and in which there isn't much time to reach a decision. Psychologist Daniel Kahneman was awarded a Nobel Prize for his work on heuristics.[4] He showed, for instance, that people employ mental shortcuts—heuristics—in many situations. The availability heuristic, for example, occurs when people utilize information that is readily retrieved from memory to make a decision. Another common mental shortcut is the representative heuristic, or the tendency to make a decision on what may superficially appear to be the most typical likely outcome, but in fact is no more probable than other outcomes.

Increasingly, we are learning that soldiers and others who make difficult decisions in situations where time is short and lives are on the line often employ intuition. In a project I worked on a few years ago we interviewed General William Scott Wallace, who, as V Corps commander, led troops into Baghdad in March 2003. When asked how he arrived at particular decisions, he at first insisted that he had relied on the MDMP. However, he eventually realized that many of his decisions were truly intuitive in nature. This is not to say they were not good decisions. A general officer with decades of experience training and fighting wars may be able to render decisions using intuition that are quite effective.

The remainder of this chapter considers how the military of the future will use cognitive psychology to better prepare its soldiers to make faster and better decisions under the difficult conditions of modern warfare. The discussion focuses on (1) battlefield cognition, (2) intuition and heuristics, (3) technology and decision making, and (4) the outcome of "cognitive dominance," or the ability to defeat the enemy through quick and agile decision making.

FUTURE DECISION-MAKING MODELS

In terms of recognizing the importance of psychological principles inherent in military decision making, the MDMP is less sophisticated than the OODA loop. The OODA loop implicitly and explicitly recognizes psychological constructs. This is partly due to the nature of an air engagement that the OODA loop was created to model, versus the deployment of ground forces that is the basis of the MDMP. However, with new technologies, evolving doctrine, and the change in the nature of the threat, decision-making theory, training, and practice must incorporate state-of-the-art psychological science to sharpen the effectiveness of future military forces.

Situational Awareness

Situational awareness is a cognitive attribute critical for making rapid and accurate decisions under stressful conditions with limited decision-making time. It represents the ability of a person to size up a situation and act in ways that maximize the chances of success and minimize the chances of failure or harm. To explain this essential attribute, I will share an anecdote from my years as a deputy sheriff in Missouri.

I was on patrol one night when a report of an armed robbery at a convenience store was broadcast over the police radio. Robberies are potentially violent crimes, and those who commit them are often desperate and ready to employ deadly force against anyone who gets in their way. I had been a deputy sheriff for over 10 years. I was intimately familiar with the area and its roads, as well as how to respond to high-priority crimes like armed robberies. The two robbers had threatened the convenience store clerk with a knife and made off with an undetermined amount of cash. The clerk maintained the presence of mind to obtain a description of the getaway vehicle and a partial license plate number. Something clicked in my mind just then: I immediately recalled there had been a series of convenience store robberies in the area over the past month, and in almost every case it was pulled off by two men displaying knives. Also in every case, they hit another convenience store within about 30 minutes.

I had to decide what to do. One option would be to turn on my red lights and siren and respond to the scene of the robbery. But other deputies were closer, and because the robbers had departed, the threat to life and limb at the location of the robbery was over. Instead, I began thinking about the location of other convenience stores relative to the scene of the robbery. This particular robbery was in a small town outside the county seat. The previous robberies had been in the county seat, a much larger city. I assumed that if these were in fact the same individuals, then they were probably from the county seat and would head back to it following the robbery. Two main roads led back to the county seat, and a convenience store was located on each road. Somehow, a vision of them robbing one of those stores popped into my head. I decided to park near one of these convenience stores in the event they did indeed rob another store.

I had barely started my stake-out when the suspect vehicle pulled into the parking lot of the second store. They did not see me. I knew it would elevate the level of danger to interrupt this second robbery (and they had not actually harmed anyone in their previous crimes), so I waited until they came out of the store and returned to their car. As they drove away, I pulled in behind them and checked the license plate. Everything matched: These were the bad guys.

Felony stops are very risky, so rather than make the stop alone, thus exposing myself and the robbers to further danger, I coordinated with other deputies by radio to rendezvous together. I waited until three backup officers were in line with my car and then I turned on my overhead lights to effect the felony stop. The suspect vehicle pulled over immediately and the two occupants complied with all demands. I arrested both subjects on charges of armed robbery, and recovered both the cash from the robberies and the knife used in the crimes. (The armed robbers were brothers—uneducated young men heavily involved in substance abuse. They were prior felony offenders, and both received extremely long prison terms of over 40 years).

This tale illustrates the concept of situational awareness. In its barest form, situational awareness is the ability of a person to (1) accurately perceive the situation in which they find themselves, (2) understand what the perceived information means, and (3) project what is likely to happen next.[5] I accurately perceived via the radio call the nature and seriousness of the crime, understood that it may have involved perpetrators who used a consistent method in their crime, and accurately projected what was about to happen next. While this may seem simple to the reader, making tactical decisions with potentially life-threatening ramifications under conditions of high stress is no simple matter.

The military began to appreciate the importance of situational awareness in the late 1990s. Its first applications, like the OODA loop, were in the domain of military aviation. Successful fighter pilots seemed to have better situational awareness skills than less successful peers. The air force partnered with psychologists to develop sophisticated models of situational awareness and implement procedures to train it. The goal was to train pilots to attend to critical parts of the battle space, build an understanding of the implications of what they observed, and to develop the skill of predicting what would occur next.

By 1998, the army realized that the concept of situational awareness formed a critical component of how ground forces, especially at the small unit level, made tactical decisions. The US Army Research Institute sponsored an Infantry Situational Awareness Workshop at Fort Benning, Georgia.[6] The workshop brought together the nation's leading situational awareness experts and partnered them with experienced infantry officers with the goal of exploring how the army could leverage situational awareness to improve decision making of infantry units ranging from small units (squads and platoons) to large units (battalions and brigades). Topics included defining, measuring, and training situational awareness for army units.

The basic model for understanding situational awareness is depicted in Figure 4.2. This shows the elements of situational awareness described above. A soldier senses the tactical environment either through direct observation, typical for small unit leaders, or indirectly through command and control displays and associated equipment, more typical for leaders and staff at higher echelons (perhaps battalions, certainly brigades and above). He must then assess the situation and quickly generate and execute decisions to address the situation at hand.

Before a soldier can make an informed decision, he must accurately assess the situation. Situational awareness is broken into three components, each of which builds on

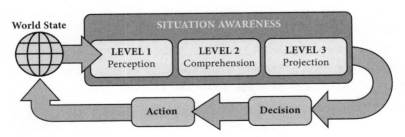

FIGURE 4.2 The Situational Awareness Model of Decision Making. (Image courtesy of Mica R. Endsley, SA Technologies.)

the preceding one. The most basic level of situational awareness is simply perceiving the environment. Dr. Mica Endsley, a pioneer in the study of situational awareness, defined this as Level I situational awareness.[7] Level II situational awareness involves comprehending or understanding the implications of what is sensed in Level I. To an experienced infantry soldier, for example, failing to see citizens in a neighborhood during a time of day when the streets and shops would ordinarily be crowded might suggest danger. Level III is the highest level of situational awareness. This level refers to the ability of the soldier to make predictions, based on what he has perceived and understood, about what is likely to occur in the near-term future. Soldiers who are sufficiently trained and experienced in reaching Level III situational awareness are in the position to then render decisions followed by actions that are more likely to be successful than those made by soldiers who are deficient in situational awareness. The outcomes of decisions made and actions taken and their effects on the battlefield then reenter the situational awareness loop and are used to generate further decisions.

Situational awareness is critical to operational success. Let's compare Lieutenant (LT) Jones, a brand-new platoon leader with no combat experience, with Lieutenant Smith, who is nearing the end of his first combat tour with respect to their situational awareness and how it may impact decision making and operational success. Both lieutenants are given orders to patrol a rural region of Afghanistan. The missions, setting, terrain, weather, and tactical situations are similar. LT Jones begins his mission and fails to perceive warning signs of possible enemy attack. Even worse, he doesn't fully understand what he does perceive. Thus unable to project what is about to occur, he and his platoon are surprised by an ambush. Hopefully his superior firepower and the tactical training of his platoon will allow them to win the fight, but things don't always work out this favorably.

In contrast, LT Smith knows where to look and what to look for, and establishes good Level I situational awareness. He quickly grasps the meaning of what he observes (Level II situational awareness), and suspects that an ambush is imminent. He immediately deploys his squads into defensive positions, sets up firing positions, and prepares to be attacked. In his case, the enemy fighters may see his preparation and abandon their plans to attack, or they may continue their plan to attack, in which case they are soundly defeated.

Talk to anyone who is involved in working in dangerous contexts and they will tell you that good soldiers, police officers, and firefighters have the ability to know what is going to happen before it happens. I have seen both sides of it in law enforcement. Most experienced officers develop situational awareness through hard experiences on the street. But some never seem to get it, or are never proficient at it. They simply don't have a clue, and they put themselves and those around them at risk because of their inability to perceive, comprehend, and project in tactical situations.

In the future, military psychologists will play a major role in improving situational awareness in two ways. First, they will develop ways to train new soldiers and leaders to develop effective situational awareness skills prior to deploying to war. Second, they will work with engineers to design military systems that augment the soldier's natural cognition with automated systems providing critical information at the right time and in the right way to establish and maintain high situational awareness at all times.

Training Situational Awareness

Situational awareness is a skill that can be both developed and trained. It can be developed through practical experience but this is time-consuming and potentially costly because mistakes will be made. In the military and other high-risk occupations, learning by doing may put both the learner and his or her associates at risk. A better strategy is to train soldiers and their leaders to improve their situational awareness skills.

Such training may occur in several ways. Certainly traditional field training contributes to the development of situational awareness. But as we saw in Chapter 3, field training is expensive and time-consuming, and therefore most individual soldiers do not receive sufficient exposure to systematically improve complex cognitive skills. There is also a classroom element to training situational awareness skills. Soldiers should know what the concept means and view having those skills as a desirable outcome. But, like riding a bicycle, hearing about situational awareness—even from the most dynamic and gifted trainer—is not the same thing as practicing those skills in a realistic venue.

The answer, of course, is that situational awareness will be trained using high-fidelity virtual simulations of tactical missions. The military has compiled countless case histories of all manner of tactical engagements, ranging from routine patrols to engaging in tense negotiations with warlords to full-scale battles. The psychology behind building situational awareness using virtual simulations is based on the understanding that quick and accurate decisions made in high-stress situations require the soldier to activate scripts and schemas and match them to the situation in which he finds himself. Briefly, a script is a pattern of behavior that matches a given situation. For example, you have a well-formed script for how to behave when entering a restaurant. You wait to be seated, expect to place drink orders, and when the drinks are delivered, you know you will be asked what you want to order for your meal. A schema is a cognitive framework that allows you to make sense of a given setting by organizing your knowledge of that situation in a meaningful way. You have a schema for a restaurant—it is a place to order food and drink and to socialize with friends—and a script for how to behave in the establishment.

Through virtual simulations, soldiers can build schemas of different tactical scenarios. By working through these scenarios and receiving feedback on their performance, they can begin to build scripts that lead to successful outcomes in given situations. To build situational awareness, a soldier should receive multiple exposures to dozens of different tactical situations. This reinforced practice builds a library of schema–script connections. As proficiency builds, the soldier can quickly pattern-match the observed situation with an appropriate script. Decision making in these cases may become virtually automatic. Observers, unaware of the prior training, may describe the soldier's decision-making ability as "instinctive," not knowing that he had experienced hundreds or perhaps thousands of similar missions prior to engaging in a real tactical mission.

This virtual training also improves Level I situational awareness by teaching the soldier what parts of the tactical environment to focus his senses on. That is, he will learn what to look and listen for, and by paying attention to the critical aspects of the mission, his perceptual skills will improve. At the same time, continued exposure to a vast array of tactical missions will teach the soldier to understand the meaning of what he senses. Thus Level II situational awareness will grow. Level III situational awareness, or the ability to predict future outcomes, comes through the ability to match the environment with scripts. The lieutenant can now anticipate an ambush and prepare his platoon to respond. The firefighter knows that a roof will soon collapse and can get his fellow firefighters out of harm's way. The police officer can sense that the suspect he is interviewing is nervous and may resist arrest, so the officer will not be caught by surprise if the suspect fights.

Augmented Situational Awareness

With modern technology, we do not have to keep all of our knowledge in our heads. We can augment cognition, including situational awareness, through a variety of information technologies. We are all familiar with a common example—remembering telephone numbers. I was leading a classroom discussion at West Point on the topic of memory and used the example of remembering telephone numbers: For example, how after looking up the telephone number of a pizza place and dialing the number, you quickly forget it. But after looking up and calling the same number several times, it gets stored in long-term memory and you no longer have to look it up. My cadets started laughing at this example and pointed out that this never happens to them. Their calls are all on cellular telephones, and the numbers they use are either programmed for speed dial or in the memory of the phone. This hit me like a club—I thought of all of the telephone numbers I had memorized in my life, and to think this was no longer necessary was a stunning revelation to me.

This illustrates, of course, that certain cognitive functions can be augmented by technology. This is most evident today in military command and control centers and in large platform weapons systems like ships and aircraft. The study of augmented cognition today is still somewhat in its infancy stages, and command and control systems employ them much the same way that we use our own computers to store and retrieve, and display information.

In the future, military augmented cognition systems will be smaller and capable of being placed on individual soldiers. They will also be better designed and programmed, matching cognitive psychology principles with the needs of soldiers to produce a fighting system (i.e., the soldier plus the systems he carries) that can not only outshoot the enemy, but outthink him as well.

A prototype of such a system for infantry soldiers emerged in the late 1990s. Known as Land Warrior, the system equipped soldiers with an array of technologies designed to enhance the ability to sense, understand, and react to the environment.[8] The Land Warrior system included a computer screen and camera so that real-time images of a mission could be shared with all members of a unit, as well as higher-echelon units. Digital maps connected with GPS information provided extremely accurate information on the soldier's location. Sophisticated radios were integrated into the soldier's helmets, allowing them to verbally communicate during missions. A heads-up display showed a range of information on friendly and enemy troops. A thermal weapon sight enhanced the ability to see under difficult conditions of darkness or smoke. Although the prototype was not fielded in large numbers, the Land Warrior concept effectively demonstrated the features that a wearable computerized command and control system could bring to a soldier.

Future systems for infantry soldiers will build on the Land Warrior concept. For infantry soldiers, weight is a critical variable. The command and control systems must be miniaturized and extremely lightweight, and long-lasting power sources are essential. Once systems are perfected, the situational awareness of the future soldier will be greatly enhanced. Thermal and other high-tech detectors will build Level I situational awareness. The soldier's computer system will help him match what he perceives to be threats likely in his particular area of operations. The intelligence and other information available at a glance will lighten the soldier's cognitive load and allow him to focus his thoughts on future outcomes (Level III situational awareness) and other higher-order decision-making tasks.

Psychologists must and will play a vital role in developing these systems. Their understanding of human cognition will help designers know what information needs to be given to the soldier and when it should be given. Only the most relevant information will be presented—overwhelming the soldier with all of the possible information that exists will only confuse him. But properly designed, future combat systems will essentially represent another brain augmenting the human brain. Each can do what it does best: The computer can use its silicon brain to receive and store vast amounts of information, keeping it available in an instant to the soldier if needed. This soldier can then use his carbon-based brain to understand, infer, and project. In short, to succeed in war we must continue to outthink the enemy, as first proposed over 60 years ago by Colonel Boyd and his OODA loop.

Intuition (Trusting your Gut)

Think about how you make decisions in life. Certainly there are many circumstances when you logically evaluate the alternatives, list the pros and cons of each option, weigh the merits of each, and arrive at a decision. For many of us, choosing a college may have involved such processes. You gathered information on the factors that mattered most to you such as location, cost, reputation of the school, possibility for financial aid, opportunities for extracurricular activities, whether you had friends attending there, and so forth. Then you ranked the schools in order of preference, and accepted an offer of admission from the institution highest on your list that admitted you. Buying a house or car, choosing among career or job opportunities, or deciding where to go on vacation are other examples. These decisions have in common the availability of information about each option and the luxury of time to arrive at a decision.

Compare this with other decisions you may have made. Perhaps you played sports in high school or college. The quarterback has at best a few seconds to size up a defense and execute a play. The basketball point guard can pass, drive, or take a jump shot, but must make that decision very quickly. A baseball outfielder decides to shade the hitter a little bit to left, hoping to get a jump on a fly ball hit in that direction. Compare these decisions with the ones mentioned above. Unlike the process of choosing a college, time is very limited. There is a lot on the line, and stress may be high. The consequences of a good decision (a touchdown, a basket, or catching a line drive) may lead to victory for your team, and a bad decision (an incomplete pass, an interception, or a line drive that is out of reach) may result in a loss. Often you may not know all of the facts. Maybe the defense will blitz, or the basketball defenders will suddenly collapse on the player you passed the ball to, or the batter may bunt instead of swinging away. Not knowing with certainty the actions of the opposing team, the decision you make may be counteracted and rendered useless or worse.

Decisions made by soldiers, police officers, firefighters, and other such personnel may be similar to each of these decision-making scenarios. The brigade commander, with his experienced and sizeable staff, may reach a logical decision on how, where, and when to deploy troops. The commander has access to a wealth of intelligence, knows the location and status of his own units and their troops, and has access to a variety of information that helps him and his staff reach a decision. The army, as discussed earlier in this chapter, calls this the MDMP. Compare this to a platoon leader on patrol in a remote area in Afghanistan. He may be the object of a surprise ambush or many other unexpected actions. If so, he cannot bring his small staff together to rationally and logically weigh out various courses of action. He finds himself much more in a situation like the quarterback. "I have to do something right now," he thinks, "or I or many of my soldiers will be killed." In the fog of battle, he may not know with certainty the enemy location, their numbers, and other information required to make a fully informed decision.

How do people make decisions like the one facing the platoon leader? Despite army doctrine, such decisions do not usually involve the MDMP, even the abbreviated one discussed in the field manual that describes military decision making. Interviews with platoon leaders (and football players, police officers, etc.) say that these decisions are often based on intuition. They may say they acted on a hunch, or that they "just knew" what to do. After the fact, they may construct an explanation of why they did what they did, but in the moment of the decision, they simply acted.

To the outside observer, such decision making may seem mysterious or represent sheer luck (if successful). But psychologists are beginning to understand the cognitive processes that make up intuition. It turns out that many of the decisions we make in life are not arrived at by the MDMP or similar processes. Heuristics and intuition play significant roles in decision making, especially in high-stakes, high-speed situations. As we shall see, our platoon leader is not making a wild guess, and this sort of decision making can be trained.

Platoon leaders and others who make fast and difficult decisions in high-risk situations often invoke intuition. The key to successful intuition is expertise. Experts, compared to novices, are faster to size up a situation and render a judgment on what to do. The science of intuition goes back to the study of expert chess players beginning in the 1940s. A Dutch psychologist and chess master, Adriaan DeGroot, studied how chess players of differing abilities approached the game.[9] He found that master chess players behaved radically differently from amateur or casual players. First, master chess players were much faster at sizing up a board. They spent less effort in understanding the location of the opposing pieces and understanding what these locations implied. They were also much better at quickly predicting what various possible moves would result in, and were able to think several moves ahead of their opponent. Psychologists refer to the ability to see the board as pattern-matching, and experts may possess up to 100,000 such patterns in their memory. By quickly assessing the pattern of pieces on the board, the chess master can then execute a good move without expending much time or laboriously evaluating the contingencies associated with each move. To the outside observer, this quick and accurate decision making appears to be based on intuition.

What is not observed is the thousands of games it takes to build the stored repertoire of positions, moves, and outcomes.

You may have already observed that the abilities of master chess players resemble the components of situational awareness. Seeing the board and the location of the pieces is akin to Level I situational awareness (perception); understanding what a particular pattern of pieces means is similar to Level II situational awareness (comprehension); and predicting the consequences of possible moves demonstrates Level III situational awareness (projection).

In the 1980s, psychologist Gary Klein began studying similar intuition-like decision making. Klein wanted to know how expert decision makers in dangerous contexts could make life-and-death decisions so rapidly under conditions that appear to a novice as pure chaos. In studying firefighters (and later, military personnel), Klein found that experts in these real life settings behaved a lot like DeGroot's chess players. A fire department chief, for example, could respond to a burning building, quickly gather information relevant to both the fire and the positioning and capabilities of his firefighters, and decide how to battle the fire. Interestingly, the expert usually arrived at a single course of action (i.e., hypothesis), despite the uncertainty of the situation and the lack of time to systematically (remember the MDMP!) arrive at and evaluate various courses of action. Klein termed this process "recognition-primed decision making (RPD)."[10] The RPD involves a feedback loop—if the decision on how to fight the fire is not effective, then the chief would quickly invoke a modification to the original plan. This cycle would continue until victims were rescued and the fire extinguished.

Klein's initial work established a substantive area of applied psychological research and application referred to as naturalistic decision making, or NDM. Its link to DeGroot's analysis of master chess players is evident. With massive amounts of experience, firefighters, law enforcement officers, military personnel, and others who make quick and difficult decisions in high-risk settings build huge mental libraries of scripts. This allows them to pattern-match a particular event with this large storehouse of information and activate a plan to deal with the situation at hand.

This understanding of decision making in these contexts makes intuition appear much less mysterious, and more importantly, implies strategies for training it. Psychologists Michael Kahana and Martin Seligman have also addressed the topic of intuition and arrived at a similar conclusion.[11] They use the example of "chicken sexing," a somewhat curious case where workers in Japan are tasked with examining newly hatched chicks and separating them on the basis of their sex. They have the uncanny ability to correctly identify the sex of the chick, but cannot articulate how they do it. It seems like magic to the untrained observer, but the ability is learned through many, many repetitions of practice.

Training Intuition in Soldiers

Meet Lieutenant Garcia. Newly commissioned from ROTC in June 2030, he attends the Infantry Officer Basic Course to learn more basic combat and leadership skills. Soon after arrival at the Basic Course, LT Garcia is taken to an immersive combat leadership trainer and taught how to use it. He is instructed to complete ten hours per week in the trainer, one of a score of similar trainers located in the virtual battle laboratory. Stepping into the trainer, the lieutenant is stepping into another world. He can move through a virtual world that feels, sounds, looks, and smells like the real world. His first missions are relatively simple and uneventful. He learns the science and art of moving subordinate troops—virtual troops imbued with the intelligence, characteristics, strengths, and weaknesses of real soldiers. In later sessions he must respond to more challenging scenarios, and all the while he receives immediate feedback on his actions. This feedback comes from both the training system itself, and in replays of his mission with expert observers who can help him understand what he did well and what he needs to improve. Over the ten weeks of the basic course, Lieutenant Garcia will execute over 500 missions, ranging from uneventful patrols, to negotiating with locals, to full-out battles with enemy troops.

Kahana and Seligman agree that intuition is "teachable, perhaps massively teachable." Teaching such skills involves both classic verbal-based instruction and, more importantly, systematically exposing the learner to a deep and broad variety of relevant situations. By providing feedback and carefully manipulating the particular circumstances of each exposure, the learner begins to build his or her own library of experiences that can ultimately be tapped into when an intuitive decision is called for. Kahana and Seligman emphasize the need for precision in creating these training scenarios: "It is crucial to closely model and overtrain 'close calls,' the scalpel-edge cases that yield the slowest response times and are most prone to error."[12]

To use Klein's naturalistic decision-making terms, our fictional lieutenant is building a sophisticated and detailed inventory of experiences—as real as those he will later build through real world missions—that will allow him to (bloodlessly, as General Scales would say) respond quickly, accurately, and effectively to the challenges he will face when his boots hit the ground in some faraway land. Combined with the traditional classroom and field training he will also receive in his pre- and postcommissioning education and training, this lieutenant and thousands like him will be better equipped than small unit leaders of today to execute missions with less threat to life (of his own unit and those of noncombatants) and less chance of making social and cultural blunders that detract from the ability of the military to achieve its larger mission goals.

Intuition can and will be trained. The technology exists to do it, and to do it on a large scale. This cannot be done without psychology. Military experts know *what* to train, but cognitive psychologists know *how* to train. Systematic exposure to evolving scenarios, immediate feedback on behavior, and guiding the transfer of learning from the virtual world to the actual world must be crafted by psychologists.

The role of psychologists in preparing soldiers for battle does not end with intuition. As we will see in Chapter 5, technical and tactical proficiency is necessary but not sufficient in preparing soldiers and their leaders to be effective in combat. They must also know how to respond, be resilient, and know how to cope with the stressors of the battlefield.

NOTES

1. Charles M. Province, *Patton's One-Minute Messages*, 65.
2. For an excellent review of the origin and development of Colonel Boyd's concept of the OODA loop, see Robert Coram, *Boyd*.
3. US Department of the Army, *Staff Organization and Operations, Field Manual 101-5*, Chapter 5.
4. See for instance Daniel Kahneman and Amos Tversky, "On the Reality of Cognitive Illusions," 582–91.
5. Mica R. Endsley, "Toward a Theory of Situation Awareness in Dynamic Systems," 32–64.
6. For a thorough discussion of situational awareness as it applies to infantry units see Scott E. Graham and Michael D. Matthews, eds., *Infantry Situation Awareness*.
7. For an infantry-specific model of situational awareness, see Mica R. Endsley, et al., *Modeling and Measuring Situation Awareness in the Infantry Operational Environment*.
8. For a description and analysis of an early Land Warrior prototype, see Jean L. Dyer, J. Reeves, and Richard L. Wampler, *Training Effectiveness Analysis (TEA) of the Land Warrior (LW) System*.
9. Adriaan DeGroot, *Thought and Choice in Chess*.
10. Gary Klein and Beth Crandall, *Recognition-Primed Decision Strategies*.
11. Martin E. P. Seligman and Michael Kahana, "Unpacking Intuition," 399–402.
12. Seligman and Kahana, "Unpacking Intuition," 399–402.

5

TOUGH HEARTS: BUILDING
RESILIENT SOLDIERS

War is for the participants a test of character: It makes bad men worse and good men better.
Joshua Lawrence Chamberlain[1]

"Touch the brain, never the same." This is a truism in cognitive neuroscience, and refers to the fact that brain damage, however minor, forever changes a person. The impact may be obvious to the casual observer, as in the case of a stroke victim who loses the ability to produce or understand speech, or more subtle changes in thoughts, emotions, or motor behavior. These effects are easy to understand because they result from physical injury that can be observed, measured, and assessed.

War also changes people. These changes can be both physical and psychological. Sometimes the effects are dramatic; other times they are hard to discern, or only manifest themselves on rare occasions. Moreover, the experience of war is not uniformly negative in nature. We frequently read and hear about physical and psychological wounds soldiers receive. Soldiers who suffer amputations, burns, or other disfigurements deserve not only all of the medical assistance we can offer them, but also our recognition and appreciation for their sacrifices. In the wars of the twenty-first century, psychological wounds including posttraumatic stress disorder (PTSD) have received an extraordinary amount of attention in the popular press. To someone not associated with the military, it may seem that PTSD is rampant among soldiers and veterans. And psychological injuries are more difficult for nonexperts to understand and relate to. There are no physical scars, nothing easy to see through casual observation. We less often hear about the resilience and personal growth that may also occur following trauma and adversity.

Psychologists pay a lot of attention to PTSD and other combat-related pathologies. I searched a national database of psychology publications and found a total of 24,004 journal articles, book chapters, convention papers, and other reports that included a discussion of PTSD. To be sure, PTSD is a significant psychological disorder. It impairs the lives of affected individuals and their family, friends, and coworkers. And many soldiers do have PTSD and suffer greatly from it. Nor does PTSD simply fade away over time. Untreated, it adversely affects veterans (and civilians with it, of course) for years. We have all seen World War II veterans, now in their 80s and 90s, interviewed in documentaries, come to tears remembering some of the events they endured during the war. Touch the brain, never the same. Experience war, never the same.

I had an incident with a Vietnam veteran when I was a young police officer in the mid-1970s that exemplifies the worst effects that war has on some soldiers. About three o'clock in the morning, my partner and I were dispatched to investigate a call. A home-owner awakened to a loud crash, looked out his bedroom window, and observed a large sedan that smashed through his privacy fence and had come to rest under his bedroom window. The motor was running and a male subject was slumped over the wheel. We arrived on the scene and observed a hole in the wood fence about the width of a 1968 Chevy Impala. We suspected that alcohol was involved in this incident, and we were not mistaken. The driver, still seated in the driver's seat, was conscious and apparently uninjured. We got him out of the car and immediately smelled the odor of alcohol on his breath. His eyes were bloodshot, and he was unsteady on his feet. In fact, he was sufficiently impaired that he was unable to perform standard field sobriety tests such as walking nine steps forward and nine back, heel to toe.

We placed the driver under arrest, arranged to have his vehicle impounded, and took him to police headquarters for further investigation. On the way to headquarters he carried on a reasonably coherent conversation and was very cooperative. He read-ily admitted that he had been drinking whisky all day, and while he didn't have a clear memory of driving through the fence, he did not dispute the fact that he had commit-ted this act. Once we reached headquarters we administered a breath test and to our amazement he registered a blood alcohol content (BAC) of 0.36 percent. This level of intoxication is equivalent to surgical anesthesia, and may be fatal for up to half of the people with a BAC this high. Nevertheless, our suspect remained talkative and could walk, albeit not in a straight line. A records check revealed that this was not his first driving while intoxicated arrest, and he been in several other alcohol related incidents with our department over the past two years. We booked him into jail and I didn't give it much more thought.

Later, I learned that this man had served in combat in Vietnam. After he returned from the army, he began drinking heavily and his relationship with his family deterio-rated. He was fired first from one job, then another. He was one of thousands of soldiers who had been exposed to Agent Orange, a defoliant widely used in the Vietnam War. He had gone to war, endured danger and deprivation, and returned a changed man.

Some months later things went from bad to worse for this veteran. He got into a violent disturbance with his wife at their residence and the police were called. I was not on duty that day, but two of our officers responded to the scene. When they arrived, this man—who had served his country in a time of war—pulled a handgun and pointed it at the officers. One of the officers, fearing for his own life, fired his service revolver, striking the man in the chest and killing him instantly. The officer had no choice—this man had committed "suicide by cop."

At that time, PTSD was not yet formally recognized as a psychiatric disorder, but this veteran showed many of the classic symptoms. His alcohol abuse, depression, inability to retain a job, and family difficulties are consistent with many of the criteria used today to diagnose PTSD. Sadly, lacking a diagnosis, he received no treatment. The small city that he lived in was a long drive from any VA hospitals. You would not have to look hard to find similar cases today with veterans of our wars in Afghanistan and Iraq. I have always felt empathy for this man, and felt that the army and society let him down.

THE PSYCHOLOGICAL REACTION TO TRAUMA MAY BE ADAPTIVE

The combination of the popular press focusing on PTSD and related disorders with psychology's almost complete concentration on these adverse consequences versus other possible sequels leads to some undesirable effects, ones that indeed can affect the combat readiness of the military. Soldiers may come to label normal reactions to trauma as pathologic, and come to believe they are mentally ill when in fact they are not. Let me explain through an illustration.

When I was in college, age 19, I became interested in law enforcement and arranged a ride-along with my hometown police department. One night I was assigned to ride with the major crimes investigator. His job was to respond to serious crimes or incidents, and his police car was equipped with specialized equipment (automatic weapons, crime scene analysis tools, life saving equipment) that other police cars usually did not have. We were dispatched to a "person down" call—in this case a person down on a residential street. Even though I was new to police work, I had already learned that such calls usually involved intoxicated individuals. But I also did not hear the radio call clearly, and as we sped to the location I noticed the officer I was with had become very tense. In sports, you would say he had put on his game face.

When we arrived at the scene, I realized why he was reacting this way. Indeed, I had not heard the call correctly. There were two officers already on the scene, standing near a body lying next to the curb. I will spare you the gruesome details, but this had been a young woman who had left a party. As she was attempting to enter the driver's side of her car, an oncoming speeding car struck her and pinned her between its left front bumper and her car, and crushed her body as it was pulled it along the length of her car. She was killed instantly. Part of her body was found several blocks away, falling from the suspect's vehicle as he fled the scene. (It did not take a seasoned detective to find the offending driver—we simply followed a gruesome trail for several blocks and found the vehicle, with some of the victim's clothing still hanging from the front of the car, parked in the driveway of a nearby home. An arrest was made. The young male driver fought us as we took him into custody—no surprise, he was intoxicated—and later was found guilty of manslaughter.)

I assisted the investigating officer with taking photos but mostly stood in shock at what I was seeing. Later, we went back to police headquarters and went through the victim's purse to see if we could find something that could identify her. The officer asked me to do this task, and I quickly found her driver's license. I could barely believe my eyes. The victim had graduated from my own high school just one year before me. She was a beautiful and popular girl. This poor young woman had been so badly mangled that I did not recognize her at the scene.

That night I had trouble sleeping. When I did sleep, I dreamed about what I had witnessed. They were vivid dreams, the kind that you are sort of aware you are dreaming, but can't stop. For several days, I felt detached from normal life. It was hard to concentrate, and images and thoughts of the disfigured remains popped into my mind, and did so beyond my control. I was irritable and preoccupied. In short, I was traumatized by the event.

Although I remember that night in great detail, the symptoms I experienced lessened with time. Within a few days, I was back to normal. In 15 years of law enforcement, I would go on to encounter many more terrible death scenes, and each time the impact was less and less. Later in this chapter, I will discuss how soldiers, law enforcement officers, and others can learn or be taught to deal with these sorts of trauma, but first I want to discuss how my reactions to this death scene relate to PTSD.

First, let's look at some of the diagnostic criteria used to define PTSD. According to the *Diagnostic and Statistical Manual of the American Psychiatric Association* (DSM), PTSD is precipitated by the experience of direct (e.g., being severely wounded) or indirect (e.g., observing an event such as a violent death) trauma that results in fear, helplessness, or horror.[2] The symptoms include recurrent and intrusive thoughts, images, memories, and dreams (beyond one's ability to control); feeling like the event is actually recurring (not recalling it in the abstract); intensive psychological distress; and physiological reactions (sweating or increased heart rate, for example). The victim shows persistent and active avoidance of the circumstances and situation associated with the trauma, persistent symptoms of increased arousal (e.g., hypervigilance, sleep disturbances, being easily startled), and significant disturbances of social, family, or work adjustment. Finally, the symptoms must persist for at least a month before being considered as representing PTSD.

With the exception of the symptoms lasting more than a month, my reactions to the death of the young woman described above map closely to the clinical symptoms that define the disorder. But here is an important point: *Not only are these reactions entirely normal, they are very adaptive.* From an evolutionary perspective, humans who react in these ways to specific trauma would be more successful in avoiding future such incidents. If my (very) far distant ancestor was suddenly attacked by a saber toothed tiger while walking along a path, it would be much to his advantage to have an enhanced memory of the event and to remain hypervigilant for some period of time following the trauma. These and other reactions would help him avoid further contacts of this kind. Maybe he would learn to avoid a particular area, never venture forth alone, or always be armed in some manner. These changes in behavior—in part motivated by the fear reactions to the initial encounter—would make him safer and therefore more likely to survive and to contribute to the gene pool.

Many if not most soldiers deployed to combat experience direct or indirect trauma. Almost all experience the natural and very human reactions similar to those I experienced, but relatively few (perhaps 15 percent) continue to experience symptoms long enough to eventually be diagnosed with PTSD or a related pathology.

Why do most soldiers eventually recover from their trauma while some do not? Some may have experienced other traumatic events before, perhaps before joining the military. Some may be predisposed through genetics or early socialization to react poorly to stress and trauma. And, of course, unlike a civilian who encounters significant trauma—who may never be placed into such a situation again—the soldier must continue to perform his or her mission and thus potentially be exposed to repeated traumatic events.

But there may also be a more pernicious factor that contributes to PTSD. The high level of attention it receives in the media combined with the attention it garners from the mental health professions may provide young and often not well-educated soldiers with a narrative that labels perfectly natural reactions to stress and trauma as being a disease. That is, they may experience a trauma, have intrusive thoughts and dreams and altered arousal, and believe they have PTSD when in fact their response is normal and adaptive.

This is not far-fetched. Even West Point cadets, who in their freshmen or plebe year all take a course in military psychology, lack an understanding of the adaptive individual responses that sometimes follow traumatic incidents. For instance, I asked West Point cadets enrolled in my cognitive psychology course to conduct a survey among their classmates that quizzed them about their knowledge of both PTSD and post-traumatic growth (PTG). We collected 100 surveys representing a good cross-section of cadets in terms of academic major and other demographics. Of this highly educated sample, 80 percent felt they had a good understanding of PTSD. In contrast, nearly the same percentage (actually, 78 percent) had never heard of PTG. And, only 22 percent of these future officers believed they "would not" or "most likely would not" develop PTSD if deployed into combat in the future.[3]

This finding is stunning. Cadets know a lot about the pathology of war, and most expect that they are vulnerable to developing such pathology in the future. And most were not aware that soldiers may also experience growth following the trauma and adversity of combat. This creates the necessary conditions for a self-fulfilling prophecy: "I have experienced a trauma, my symptoms are those of PTSD, therefore I have PTSD."

If this is true among highly educated and motivated West Point cadets, it may well be the case that less educated soldiers with equal exposure to the media coverage of PTSD but less exposure to the science of psychology may be even more at risk to label their reactions to combat trauma as a sickness. Moreover, most of their officers and senior noncommissioned officers do not understand the psychology of stress and adversity, and may either be dismissive of the concept of PTSD entirely, or subtly reinforce self-labeling because it is the only narrative they know as well. Neither outcome is desirable.

A FULL SPECTRUM OF RESPONSE TO COMBAT

General Chamberlain was right; the news is not all bad. Adversity and sometimes trauma may lead to positive personal growth. Growing from trauma and adversity does not make headlines, nor until recently has it been extensively studied by psychologists. Recall that a simple search revealed over 24,000 published scientific papers on the topic of PTSD. A similar search on the topic of posttraumatic growth (PTG) yielded only 1,064 papers. At the rate of almost 25 to 1, psychologists study the pathological consequences of exposure to trauma over more favorable outcomes.

But what is PTG, exactly? Studies of victims of violent crimes, cancer survivors, and others who have endured and survived life-threatening trauma clearly indicate that in some cases—perhaps in the majority of cases—the survivor feels that he or she is a better person in some way. Some may come to value their family more; others make time to savor life; still others may find a deeper meaning and purpose in life through religion or a sense of connection to humanity.[4]

It is important to understand that, by definition, PTG requires the experience of trauma. The same types of trauma that may be followed by PTSD may also result in growth. In addition, the reader should understand that nobody is advocating the experience of significant trauma as a basis for personal growth. And it is also important to recognize that a person's response to trauma is highly individualistic, and influenced by a host of personal, social, and contextual factors. When trauma occurs, we may experience mostly negative reactions, perhaps mostly positive reactions (at least, over time), or, more likely, a mixture of both negative and positive reactions to the event. Being raped is a horrific ordeal that results in a host of negative reactions including anxiety, fear, depression, and often PTSD symptoms. But that same victim may also show evidence of personal growth. Trisha Meili, who barely survived a brutal rape in New York's Central Park in 1989, later used her experience as the basis for outreach to help other victims of violence.[5]

To complicate things further, there appears to be an intermediate set of reactions to adversity and trauma that are not pathologic or representative of growth. Resilience, or the ability to bounce back from stress, represents this middle ground. A resilient person may experience trauma and rather quickly return to normal functioning in his or her life. If defined as a return to baseline functioning, then someone who is functioning at a very low level prior to trauma may simply return to or remain at that level following the trauma. Or baseline functioning may be so high that such a highly resilient person may not have room to improve functioning further. Unfortunately, there are no widely accepted definitions of resilience and PTG, but in any case they represent nonpathologic responses to trauma.

Both PTSD and PTG involve the experience of trauma. Most often trauma is a discrete event that may only occur once in a person's life. A victim is shot in a robbery, or a driver witnesses the death of another person in a motor vehicle accident. Clearly such events may precipitate both PTSD and PTG. But what about other situations that, by their very nature, test a person's resolve, emotional stamina, and ability to cope, but are not tied to a specific trauma as defined in the DSM? I have found that soldiers often develop profound psychological responses to combat even in cases where they may not have been personally exposed to trauma. Being separated from one's family for 12 months during a combat deployment may pose tremendous challenges to a soldier. Combat deployments are characterized by short and irregular sleep, social isolation, extreme temperatures and other weather, and physically taxing work conditions. The deployed soldier may find him or herself in a poorly led or organized unit and be constantly frustrated by a poor organizational climate.

Perhaps a better word for such situations is "adversity." Adversity may include trauma, but extends to other significant challenges of the sort described above. When understanding the personal dynamics of combat deployments, adversity may be a more descriptive term. Not all soldiers experience trauma, but all experience adversity in one form or another. So for the purposes of understanding stress, resilience, and growth in the military context, substituting the word adversity for trauma makes sense.

This is not a trivial distinction. A colleague of mine in another academic department was deployed to assist in the Iraq War effort a few years ago. He held a doctorate in philosophy and was a respected teacher, scholar, and colleague. West Point deploys its senior military faculty to operational settings periodically to keep them "green"—or, put another way, to help them maintain their military knowledge and expertise. During his deployment, this officer found himself in a situation that he felt to be unethical and illegal with respect to the use of funds to support the Iraqi military and community. Despite reporting the misuse of funds, nothing was done. He felt frustrated, pressured to conform, and angry, and must have felt he had nowhere to turn. Unfortunately, this officer chose to end his life with a self-inflicted gunshot wound. He did not face trauma in the traditional sense, but he did face severe adversity. And consequently, his family lost a loving husband and father, and the army lost a brilliant and talented officer. And this is not an isolated case. In a longitudinal study of PTG that I am conducting, I recently interviewed a field grade officer who worked in a unit led by an unethical officer who asked his soldiers to perform illegal acts. It is difficult to change units once deployed, and the morale of that unit was broken. This officer considered suicide while stuck in that situation and, now, over a year after leaving this unit, is still depressed, has difficulty relating to his spouse, and drinks heavily.

While the vast majority of research into the impact of combat on soldiers focuses on negative outcomes, there is growing evidence gathered by psychologists that war may provide a source of growth as well. You may be surprised to learn that even for units that experience the harshest combat conditions, the highest PTSD rates appear to be around 30 percent. In other words, well over half of these soldiers do *not* experience diagnosable pathology. This seems to be true in other highly stressful situations, such as being a prisoner of war.

And these are the highest rates observed. More typically, PTSD rates top out at around 15 percent. And not all types of military units are affected the same way. Highly trained and specialized units including SEAL teams, Rangers, and other elite organizations show very low (less than 5 percent) PTSD rates, despite being involved in frequent and intense combat. Contrary to what you may believe, based on extensive press coverage and professional attention paid to PTSD, it is the least frequent response to combat.[6]

What about more positive responses? An early study of the long-term effects of combat exposure on soldier adjustment examined over 1,200 soldiers who had served in World War II, Korea, and Vietnam.[7] This landmark study looked for both negative and positive effects of war on subsequent adjustment. For these combat veterans, the positive effects of war were more common than negative ones. Positive effects included mastery, self-esteem, and coping skills. Moreover, these effects were more pronounced among those who had served the most months in combat, and served to mitigate negative effects such as depression and PTSD.

Other psychologists have found that among Americans held as prisoners during the Vietnam War, over 60 percent believed they had experienced personal growth, despite having been held captive, tortured, and often wounded badly prior to being captured. Similar to the study mentioned above, the more severe their experiences, the greater the personal benefits they felt they had derived from their ordeals. Studies that directly assessed PTG support the notion that trauma and adversity may lead to growth. About two-thirds of a sample of aviators shot down and held captive in Vietnam, when given a psychological test that measures PTG, showed evidence of growth. This growth was particularly evident in personal strength and appreciation of life.[8]

I conducted a study of officers who served in the Iraq and Afghanistan wars.[9] Over 100 army officers completed the survey, given to senior captains and junior majors who had recently returned from combat tours. All had completed one such tour, and most had completed two or more. They had experienced the most severe challenges of combat. I asked them to rate 24 personal character strengths and to indicate whether these strengths had decreased, stayed the same, or increased following their combat experiences. These officers indicated the highest growth in the following five personal character strengths: teamwork, capacity to love and be loved, bravery, gratitude, and honesty. Perhaps these strengths were reinforced by their combat experiences. When asked to rate the importance of each of the 24 strengths to their ability to cope effectively with a major challenge encountered in their last combat deployment, they rated four of these same strengths—teamwork, bravery, capacity to love and be loved, and honesty—as being crucial to responding to that challenge. Persistence replaced gratitude on this list.

A MODEL OF PERSONAL ADJUSTMENT TO COMBAT

It is clear that human beings who experience severe trauma and adversity respond in a wide variety of ways. Some are debilitated, others remain relatively unchanged, and still others grow in important ways. Future efforts to train, educate, and develop soldiers must rely on a more sophisticated model of the effects of combat than a simple dichotomy of "sick" (e.g., PTSD) or "well" (the absence of diagnosed pathology).

It is useful to think of the effects of combat on soldiers as a range along a continuum. The curve shown in Figure 5.1 is one way to illustrate this continuum. You can use this curve to illustrate the general classes of reactions to combat. Psychopathologies that are clinically diagnosable are to the left of the first cut-off noted on the left side of the curve. These may represent about 15 percent of the reactions that commonly follow combat exposure. On the far right of the curve, another line is drawn. To the right of this line, we may represent positive outcomes that follow combat exposure, including PTG. This too may be seen as representing 15 percent of soldier reactions to combat. This leaves 70 percent between these two extremes, and may represent varying degrees of resilience.

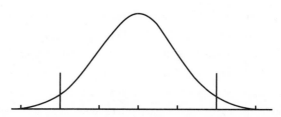

FIGURE 5.1 Psychologists traditionally focus their efforts on the far left end of the normal curve, reflecting pathology. Future psychologists will expand the focus to the right side of the curve, extending their focus to resilience and flourishing. (Image courtesy of author.)

A close look at Figure 5.1 reveals some interesting implications. It is clear that PTSD, for instance, is not something you have or do not have, with no middle ground. Based on established clinical criteria, soldiers in the left-most portion of the curve may be diagnosed with PTSD. But what about those who are just above the cut-off line, or within a fairly close distance? Perhaps they have many symptoms of PTSD, but not enough to warrant a clinical diagnosis. For the sake of argument, let's say that this gray area would encompass an additional 10 percent of soldiers who deployed to combat. These people may have serious problems in personal, social, and family adjustment but not sufficient to result in a formal clinical diagnosis that would in turn warrant treatment either in military or veterans health care systems. The number of such cases is not trivial. Over 2 million soldiers have served in Iraq and Afghanistan. Ten percent would equal 200,000 cases of impairment that are not officially recognized by the health care system. What is being done to help these people? Added to the 15 percent that are diagnosed, this would be a half-million men and women who suffer some degree of impairment or pathology stemming from combat exposure.

The far right of the curve is also interesting to consider. Those who are in the far right may represent true PTG. But those near the cut-off may also have shown important personal growth in some dimensions. Following the same argument as above, if the 10 percent that approach the PTG cut-off are really doing well, then we also have in the neighborhood of a half-million soldiers and veterans who are flourishing following combat.

That leaves the middle 50 percent of the curve. To one degree or another, we can think of this as the "resilient zone." The 1 million soldiers and veterans who fall into this zone either were relatively unaffected by their experiences or quickly recovered back to normal functioning. Like my father following World War II, they return to their normal military duties or to the civilian world and go about the everyday business of work, raising families, and being a part of a community.

Of course, no simple diagram can completely capture all the nuances of the impact of combat on human adjustment. People who have experienced combat may migrate along this curve depending on the situations they find themselves in at a given time. So a person may devolve under life stress and later be diagnosed with PTSD. Another may move from resilience to growth. Paradoxically, as I discussed earlier, it may be possible to simultaneously occupy more than one position on the curve. Just as an amputee who is debilitated to some degree from the loss of a leg may also embrace the importance of family or have developed stronger personal character from her experience, someone with negative symptoms of hypervigilance or sleep disorders may too show some signs of personal growth in other aspects of their adjustment.

Figure 5.1 also illustrates the important notion that future programs aimed at improving the adjustment and resilience of soldiers must focus not just on the far left of the curve, where efforts emphasize treating pathology. They must also provide training, education, and developmental experiences that allow all soldiers to move to the right on this curve. I suspect that the greatest potential for a positive impact may be on those to the left of the middle of the curve. Even modest improvements in perceived self-efficacy, confidence, and coping skills may translate not only into tangible improvement in quality of life for the soldiers, but also will have positive organizational consequences to the military in terms of increased morale, combat readiness, task performance, and retention.

BUILDING TOUGH HEARTS IN TODAY'S MILITARY

The future military will be extremely technical, very expensive to maintain, and small. Maintaining large twentieth-century cold war armies in the future will be neither practical, given the types of threats likely to occur, nor affordable, given the cost.

Because of these constraints, each individual soldier will matter more than ever to the combat capability and readiness of future forces. Nations will call on their militaries to serve in a wide array of tasks, ranging from peacekeeping and nation-building, to intervening in regional conflicts, to full-scale war. In short, future war will be every bit as stressful and challenging as current war.

In this context, maintaining the mental and emotional health of soldiers will be critical to maintaining force readiness. An army of 500,000 soldiers cannot afford to have 10 percent of its members functioning at less than full capacity because of mental disorders. As I write this paragraph, the United States has about 90,000 military personnel deployed in Afghanistan. It is not hard to imagine the impact of a high rate of psychological injuries on unit effectiveness. Even a rate of 5 percent translates into 4,500 soldiers who are not combat ready or function at low levels of effectiveness.

Historically, the military has invoked two strategies for dealing with psychological injuries. The first was to simply replace the affected soldier with a new recruit. Up through the Vietnam War, the United States fielded a conscripted military. If the army needed more soldiers, it could just draft more young men, give them the necessary minimum training, and put them into operational units. The all-volunteer-force that replaced the conscripted military is much smaller and depends on attracting sufficient numbers of qualified recruits to fill its ranks. This is no small task, even for the much-reduced military of the twenty-first century. In the United States today, only one in four people between the ages of 17 and 24 even qualifies to serve in the military.[10] Some have criminal records, others do not meet minimal standards on the ASVAB, and an increasingly high number are incapable of performing military physical fitness requirements due to obesity. So replacing incapacitated soldiers through recruitment is a daunting challenge, compounded by the longer training needed in the twenty-first century to prepare new soldiers to operate the more complex weapons and other equipment that characterize the modern military.

The second response was to treat psychological injuries after they occur in hopes of returning soldiers to the battlefield. This approach is based on the disease or medical model of pathology. A wound occurs, you treat it, and the soldier is then either restored to effective functioning or is disabled and removed from duty permanently or until such time they recover sufficiently to resume their duties. This disease model emphasizes treatment rather than prevention and is reactive rather than proactive.

Until 2008, the US military used the disease model as its primary response to psychological injuries. A psychological problem was noted and diagnosed, and then a medical treatment regimen was invoked. This approach has many flaws, including the fact that it stigmatizes soldiers to the point where many—perhaps most—would not seek treatment for psychological issues. And once a pathology becomes too great to be ignored, it is often too difficult to treat effectively.

At the same time, the toll of war was beginning to overwhelm the ability of the military health care system to treat the influx of psychological wounds. They simply could not hire enough psychologists, psychiatrists, counselors, and social workers to respond to the flood of soldiers returning from Iraq and Afghanistan with significant psychological problems. In short, the disease model was overwhelmed and was incapable of responding to the sheer numbers of psychological casualties that were piling up.

It was in this context that, in 2008, General George Casey, the Chief of Staff of the US Army, recognized that a reactive, treatment-based approach was not enough to care for soldiers affected by combat related stress disorders. General Casey called together a team of experts to plot a strategy for the army to change the way it did business in dealing with psychological casualties. This meeting would not have occurred without the efforts of Army Colonel Jill Chambers. Colonel Chambers had begun working on the general problem of soldier resilience for Admiral Michael Mullen, then the Chairman of the Joint Chiefs of Staff. General Casey caught wind of this effort, and asked Colonel Chambers to assist him in organizing his team of experts. Colonel Chambers worked tirelessly to promote the importance of psychology in improving soldier adjustment and performance, and skillfully navigated through sometimes difficult political waters to ensure this meeting took place. In no small measure, hundreds of thousands of soldiers benefited by Colonel Chambers's passion and dedication to inspire the army's senior leadership to address the emotional and psychological adjustment of its soldiers.

Based on General Casey's initiative, the army founded a program called Comprehensive Soldier Fitness (CSF). General Casey tasked Brigadier General Rhonda Cornum, an army physician, to stand up the CSF program. He could not have made a better choice. General Cornum was not just an excellent surgeon; she was also a decorated combat veteran. You may remember that she was the American Army doctor whose helicopter was shot down in the first days of Operation Desert Storm in 1991. She was taken captive by the Iraqi Army and held as a POW. Severely wounded, she survived her ordeal and, after repatriation, continued to serve as an army officer until 2011. I recommend the book *She Went to War: The Rhonda Cornum Story* that tells the story of her mission, capture, POW experience, and its aftermath.[11]

Soon after founding CSF, General Casey invited Dr. Martin Seligman, Dr. Rich Carmona, Dr. Larry Dewey, and me to a lunch meeting to discuss and plan the scope and fundamental approach to the emerging CSF program. Comprehensive Soldier Fitness was conceived as a proactive, training-based approach to building mental and emotional health skills in soldiers *before* they are exposed to combat. Its goal was to *prevent* psychological injuries, not simply *treat* them. It was to be run by the operational side of the army, not the medical side, and was decidedly not based on the disease model.

Dr. Seligman is the founder of positive psychology, a new area of psychology that focuses on a scientific understanding of adaptive and effective behavior. Positive psychology is interested in what allows people to flourish and grow, and aims to understand the psychological processes that underlie the ability of humans to prosper psychologically even in the face of extreme challenge. In 2008 I had just completed a year-long term as president of the Society for Military Psychology, and my presidential theme was bringing positive psychology applications into military contexts. Dr. Carmona is the former surgeon general of the United States, and has long championed an overhaul of the disease model of health care. Dr. Larry Dewey is a veteran psychiatrist employed by the Department of Veterans Affairs (VA), with three decades of experience treating psychological wounds incurred in war. Figure 5.2 is a photo of this initial CSF meeting.

FIGURE 5.2 Meeting with General George Casey, Army Chief of Staff, Nov. 25, 2008. (Photo courtesy of D. Myles Cullen.)

This lunch meeting with General Casey was pivotal in shaping the future of CSF. Dr. Seligman and I argued that the military is a natural home for positive psychology. It consists mainly of young people who are physically healthy, and has a strong culture that supports personal growth. As a high school student, I remember the military being touted as a way of building strength and maturity. Many of my friends joined the military after high school and benefited greatly from it. The military culture believes that anyone can improve and grow, and prides itself on providing training, schools, and an organizational climate that promotes personal growth.

General Casey and his staff (which included the army surgeon general, a retired four-star general, and several other senior army officers) quickly seized on the idea that positive psychology could serve as the scientific cornerstone of CSF. Dr. Seligman described empirically validated approaches that have been used in other contexts to build resilience and provide people with the personal and social skills needed to flourish, rather than languish. In short order, a plan was built to flesh out the details of CSF and to provide the army with a plan to create a more resilient army. The details of the early development of CSF can be found in a special issue of the *American Psychologist*, coedited by Dr. Seligman and myself.[12]

I felt a bit like Forrest Gump at that meeting. Although I had argued repeatedly that positive psychology was a perfect match for the military, it was Dr. Seligman who had the extensive conceptual and empirical knowledge to describe for General Casey what CSF should become. What was very clear to me, however, was what a unique circumstance it was for the military to turn to psychology as an answer to one of its most pressing issues. In 30 years as a military psychologist, I had time and again seen programs based on sound psychological research and principles ignored by the senior leadership of the military.

Moreover, the time was right for psychology to play a role. Positive psychology had been founded 10 years before, and in that short time had generated significant developments in psychology's understanding of positive human adaptation and performance. If psychology had been called upon in the 1990s, I don't think it would have had the tools to help the army shape a CSF-like approach to building soldier psychological fitness. For psychology as a discipline, it was a classic case of being at the right place at the right time.

Based on its positive psychology roots, CSF focuses on giving soldiers the personal and social skills needed to be more resistant to the rigors of combat and combat deployments, to be more resilient in the face of adversity, and to learn to grow in response to trauma and adversity. The focus is on encouraging personal and social growth, not on treating pathology. This focus is consistent with a strong military culture that embraces the notion that training and education can provide soldiers with the tools needed to succeed. Because the focus is on training and education with the intent of improving the combat readiness of the army, organizationally CSF was placed under the supervision of the operational army, not the Medical Services Corps. This is an important decision. Defining emotional fitness as a training and education issue reinforces the notion that the programs offered under the umbrella of CSF are not designed to treat sick or otherwise deviant soldiers, and helps avoid the stigma associated with treatment. The fact that every soldier in the army is required to participate in CSF also underscores its training orientation.

The CSF program can be looked at as akin to the physical fitness training that has long been a part of military culture. The army and other branches of the military have well established physical fitness standards coupled with regular physical fitness assessments. If a soldier is deficient in one or more components of physical fitness (a two-mile run, sit-ups, and push-ups), there are various resources to help the soldier improve. For example, all units have a master physical fitness trainer, an NCO who has attended a special training program to prepare them to teach, train, and develop physical fitness skills in others.

The cornerstone to CSF is assessing resilience. If you can't measure it, you can't improve it. A small group of prominent psychologists was brought together at the University of Pennsylvania under the direction of Dr. Seligman to define resilience and to develop a simple but psychometrically sound metric to assess it. Early on it was decided to substitute the word "fitness" for resilience. Ordinary soldiers understand what fitness means but may find the word resilience confusing at best. After much discussion, it was decided that psychological fitness could be broken down into four domains: emotional, social, family, and spiritual fitness. The use of the term "spiritual fitness" was roundly debated. Most of the psychologists on the team felt that this part of fitness deals with meaning and purpose, and thought it should be called that. The military consultants felt it better to use the word spiritual because of the long-standing role of chaplains in the army who have historically been a resource for soldiers seeking help for psychological problems. It should be noted that in the military, chaplains perform an important role in counseling and providing advice independent of their religious duties, and are often a trusted source even for soldiers who do not have strong religious beliefs.

Once the four areas of psychological fitness were defined, then the team set about developing a single measure that would assess each of the components. The goal was to incorporate as many existing psychometric tests as possible and only to use tests or questions known to be reliable and valid. The metric needed to be relatively short and written in a language that all soldiers, regardless of educational background, could easily understand. What emerged was the Global Assessment Tool (GAT). The GAT contains just over 100 items and generates a fitness score for each of the four components.[13]

Following a short period of preliminary testing, the GAT was placed on an army website and all soldiers were directed to complete it. Within weeks, several hundred thousand soldiers had completed the GAT. From these data, CSF psychologists were able to begin assessing both the reliability and validity of the GAT. Over time, questions are added and deleted with the goal of including only those questions that provide the most information about psychological fitness. When soldiers complete the GAT, they immediately see their scores expressed in a normative manner. That is, they can see in what areas of fitness they are at, above, or below average. For those who appear to be low in one or more of the fitness categories, suggestions for improving in the appropriate area are given.

It is important to note that only the soldier completing the GAT sees his or her score. No commander or supervisor, or physician or psychologist has access to an individual's GAT scores. Without this restriction, soldiers might be reticent to answer questions honestly. So at the individual level, the GAT is aimed only at individual growth and development.

It is possible, however, to aggregate GAT scores at the unit level. This preserves the anonymity of individual responses but provides unit commanders with an important assessment tool. For example, suppose that prior to a combat deployment a brigade commander looks at the average GAT scores for each of his subordinate units. He finds that one battalion is very low in emotional fitness. Knowing this, the commander can initiate unit training to increase emotional fitness in that specific unit. Or, following a combat deployment, the commander can quickly get a read on the emotional, social, family, and spiritual fitness of all of his subordinate units, and launch strategies to build or rebuild fitness as needed.

At the same time the GAT was being developed, various strategies for training emotional, social, family, and spiritual fitness were explored. Psychologists who were experts in each of these domains met with army psychologists and leaders to develop methods that were both effective and congruent with army culture. The ultimate goal is to develop a variety of training protocols for each domain that include individual and unit training in various formats.

Let me illustrate with emotional fitness.[14] Emotional fitness involves many skills that can be taught and learned. The problem was how to get the training in an organization of over a half-million members. We decided to implement a train-the-trainer plan, modeled after army master physical fitness trainers. The Penn Resilience Program (PRP) was used as the model. The PRP, originally designed to train teachers to develop resilience skills in their students, was adapted for application to the army. Instead of training teachers, mid-level NCOs were selected to attend a 10-day training program conducted at the University of Pennsylvania. This program teaches recipients to develop resilience skills and mental toughness, to identify personal character strengths, and to build and strengthen social relationships. Additional training is given on how to sustain and enhance psychological fitness. Over 10,000 NCOs have completed the master resilience training (MRT) course to date. It is now being expanded to Fort Jackson, South Carolina, and other army installations, and is quickly helping transform the culture of the army to embrace psychological fitness, as well as providing all soldiers in the army the opportunity to build these skills.

Parallel efforts have been implemented for social, family, and spiritual fitness.[15] Some training programs are available online, others involve group instruction. Training in all aspects of psychological fitness is being conducted for new recruits through mid-level and senior service schools. This is not a one-time training event, but a continuing effort to train and develop fitness skills in all soldiers through their entire army careers.

Initial validation studies show that the GAT is a reliable and valid assessment tool.[16] Soldiers who score high on the GAT are more productive, more likely to remain in the army, and less likely to develop psychopathology. Those who score low are at higher risk for pathology, substance abuse, conduct issues, and suicide. Early empirical evidence also shows that the training methods developed under CSF are proving effective in increasing psychological fitness.

BUILDING TOUGH HEARTS IN TOMORROW'S MILITARY

The relevance and importance of psychology and allied disciplines in selecting, training, and developing psychologically hardy soldiers will expand exponentially over the next decades. These developments will supplement advances in the understanding of combat-related pathology, and will combine into a full-spectrum approach capable of aiding soldiers throughout the curve shown in Figure 5.1. I foresee three overlapping areas of how psychologists can enhance the psychological body armor of future soldiers.

Incremental Development of CSF-like Programs

The CSF program will show evolutionary growth through the addition of new and more effective resilience assessment and training methods. Ineffective components will be dropped. These programs will be adopted by all of the US military services and tailored to the unique needs of each service. The types of combat stress experienced by air force drone pilots, who employ highly lethal weapons but do not view the details of their effects on enemy soldiers, will necessitate resilience training skills that may differ from those needed by ground troops, who experience trauma firsthand.

Programs similar to CSF will also be adopted by the military organization of other nations. I recently gave a keynote speech on CSF at a military psychology symposium sponsored by the Indian Ministry of Defence in Delhi. The Indian army expressed deep interest in developing a resilience training program to help its soldiers adapt to the unique challenges they face. It is fair to note that the Indian army faces some threats not faced by the United States that may also require specialized resilience skills. Unlike the United States, the Indian army must use military force against insurgents within its own country. Employing military power against fellow citizens can be uniquely stressful.

By the year 2030, no soldiers in the US Army (with the exception of just a handful of senior general officers) will be able to remember a time when resilience training was not a significant part of their training and development, just as no current soldier can remember a time when physical fitness training was not an essential part of the experience of being a soldier. This will virtually eliminate the stigma associated with psychology, psychologists, and seeking help for psychological problems. If a soldier injures her knee, she sees a physician and perhaps a physical therapist to heal and strengthen it. If she finds herself unable to concentrate, irritable, depressed, or otherwise suffering from psychological pain, she will seek help from psychologists and other mental health experts. All soldiers will want to improve their resilience. In army culture, one's physical fitness training score (based on a PT test taken twice a year) is a point of pride. With improved metrics, perhaps the soldier of the future will value and even take pride in improving their resilience score.

Resilience Training Will Become Increasingly Part of Integrated Simulation-Based Training

Future training, as we saw in Chapter 4, will increasingly be delivered via highly advanced and very realistic simulations. The same simulations used to hone tactical decision making, negotiation skills, and cultural competency will be employed to strengthen personal and interpersonal resilience skills.

It is highly artificial to separate the training of cognitive skills from those skills needed to maintain emotional and psychological health. Bloodless learning involving progressive exposure to combat conditions can shape both skill sets. As the soldier builds his mental library of battle scripts, he also can begin to normalize reactions to killing, death, dying, and other stressful components of combat. Future combat simulators will replicate real combat with high fidelity across all five senses. The soldier will experience neural and endocrine system arousal associated with combat, and learn how to cope with it during the fight and its immediate after effects. The reader may see this as deliberately desensitizing soldiers to the brutal realities of war, maybe even making them more callous about killing. They may be right. Soldiers do not have the luxury of mulling over whether to kill or not, nor can they simply quit if they find circumstances to be overly stressful. In fact, when presented with a threat, they must decide within seconds or fractions of seconds whether to employ force. Not to do so endangers their own lives and those of their fellow soldiers.

As important as learning the cognitive and emotional components of combat are, it is equally important to use the simulations to train soldiers on how to deal with post-traumatic stress. This can emphasize short-term strategies most applicable just after a fight such as relaxation techniques, as well as long-term strategies like meaning-making and sense-making.

It may be possible to develop simulation-based procedures to ready soldiers for their return to civilian society. It is unjust to both the individual soldier and to the society to which they will return to not have such training. I know of no systematic programs that currently achieve this. Although soldiers are given some time to adjust following their combat tours before returning to their families, this amounts most often to completing administrative duties, rest, and simply time to unwind.

This is not enough. History has taught over and over that returning to civilian life following combat is not something that comes easily. Soldiers often have difficulty readjusting to domestic life or routine jobs. They may argue or fight with their spouse and children, or have difficulty keeping a job. They may turn to alcohol or other mal-adaptive means of dealing with stress.

Future resilience training, both simulation-based and live, must develop systematic, scientifically based methods for providing the resilience skills needed to make the transition back to civilian life. Besides helping soldiers and their families, the economic impact on society in terms of decreased health care costs coupled with increased productivity of the soldier will be immense. It takes months or more to train a person to be an effective soldier, so why do we think that we can simply discharge them from the military at the end of their service and expect them to make a seamless transition back to ordinary life?

These resocialization training programs, aimed at changing a tough soldier's heart back into a warm civilian's heart, could be the greatest contribution that psychology will make later in the twenty-first century. The science is yet to be developed, but the principles of positive psychology—with its focus on human achievement and flourishing—may provide the place to start.

Behavioral Neuroscience Will Develop Methods to Maintain Resilience

Traditional psychologists will be aided by their colleagues in neuroscience in developing interventions to improve, sustain, and sometimes restore resilience. Advances in neuroimaging techniques are already allowing scientists to look into healthy, living brains and to isolate brain structures and systems that underlie human behavior. Once these structures and systems are identified, it may be possible to develop drugs that can either create a positive, resilient reaction to traumatic stress and adversity and/or suppress negative, pathologic reactions.

Stress is already known to affect the function of brain areas important in the forming of traumatic memories. Scientists have identified a host of neurohormones that are associated both with negative and positive stress responses. We are beginning to understand how a brain predisposed to a negative reaction to traumatic stress may differ in its neural and hormonal response compared to a resilient brain. As the nuances of the neuroanatomy and neurochemistry of PTSD, resilience, and PTG progress, this raises the possibility that brain-based interventions may be developed to create positive, adaptive responses.

These techniques could involve active stimulation of selected brain structures via miniature, wireless electrodes. Drug-based interventions are possible. Drugs that alter the brain's hormonal response to stress might act as a prophylactic against PTSD. Sleep and arousal patterns, also important correlates of combat stress, may be similarly addressed.

Of course, any neuroscientist knows that the brain is a very complicated organ and even tiny interventions in one area will often have unexpected effects in other areas or functions. We do not presently have the knowledge to create a resilient brain by way of these sorts of interventions. But this is a book about future possibilities, not current capabilities. The requisite knowledge may arrive sooner versus later, and the military may soon be leveraging neuroscience as discussed here.

Perhaps more so than in any other topic of this book, the prospects of using brain science to build better soldiers raises significant ethical issues. Ignoring the science will not make these issues go away. As we will see in an upcoming chapter, science is already laying the groundwork for developing super soldiers who represent a blend of natural human and engineered capabilities. It will be fascinating to watch this story unfold.

NOTES

1. Chamberlain, Joshua Lawrence, *The Passing of Armies*, 295.
2. American Psychiatric Association, *Diagnostic and Statistical Manual of Mental Disorders DSM-IV-TR*, 4th ed., 463–68.
3. Michael D. Matthews, "Self-Reported Knowledge of Posttraumatic Stress Disorder and Posttraumatic Growth Among West Point Cadets."
4. Richard G. Tedeschi and Richard J. McNally, "Can We Facilitate Posttraumatic Growth in Combat Veterans?" 19–24.
5. Trisha Meili, *I Am the Central Park Jogger*.
6. For a thorough review of stress and resilience in a military context, refer to Brian J. Lukey and Victoria Tepe, *Biobehavioral Resilience to Stress*.
7. Carolyn M. Aldwin, Michael R. Levenson, and Avron Spiro, "Vulnerability and Resilience to Combat Exposure," 34–44.
8. Adriana Feder et al., "Posttraumatic Growth in Former Vietnam Prisoners of War," 359–70.
9. Michael D. Matthews, "Character Strengths and Post-Adversity Growth in Combat Leaders."
10. William Christeson, Amy Dawson Taggart, and Soren Messner-Zidell, *Ready, Willing, and Unable to Serve*.
11. Rhonda Cornum and Peter Copeland, *She Went to War*.
12. Martin E. P. Seligman and Michael D. Matthews, eds., *American Psychologist Special Issue*.
13. Christopher Peterson, Nansook Park, and Carl A. Castro, "Assessment for the US Army Comprehensive Soldier Fitness Program," 10–18.
14. Sara B. Algoe, and Barbara L. Fredrickson, "Emotional Fitness and the Movement of Affective Science from Lab to Field," 35–42.
15. John T. Cacioppo, Harry T. Reis, and Alex J. Zautra, "Social Resilience," 43–51; John M. Gottman, Julie S. Gottman, and Christopher L. Atkins, "The Comprehensive Soldier Fitness Program," 52–57; Kenneth I. Pargament and Patrick J. Sweeney, "Building Spiritual Fitness in the Army," 58–64.
16. Paul B. Lester et al., *The Comprehensive Soldier Fitness Program Evaluation, Report #3*.

6

WINNING HEARTS AND MINDS

If the people are against us, we cannot be successful. If the people view us as occupiers and the enemy, we can't be successful and our casualties will go up dramatically.
General Stanley McChrystal, *Afghanistan 2009*[1]

Those of us of a certain age will remember watching the evening news during the Vietnam War, and hearing the daily and weekly body count of dead US service personnel. The venerable Walter Cronkite would end his broadcast each Friday with the number of American soldiers killed the past week. Field commanders emphasized the importance of "search and destroy" missions aimed at killing as many enemy combatants as possible. This focus on body count as a metric for military success represented a significant change from both world wars, where success and failure was measured by territory gained and lost. Arguably, the American public did not understand this index of military success. Counting bodies in the absence of geographical and political gain, with no end to the war on the horizon, may have felt like a football game where the scorekeeper counted tackles, but not yards gained or points scored. It was not congruent with American culture, where winning and losing are defined based on gains and outcome, not simply on process.

Another ethnocentric trait of Americans was incongruent with body count as a valid index of military success. At the time of the Vietnam War, America had emerged as the world's dominant nuclear superpower. We had a vibrant economy and we were winning the space race. In short, Americans believed they were superior to citizens of other countries, especially when compared to a country such as Vietnam that was so far removed from western culture, religion, and values, had relatively little economic clout, and that many Americans had not heard of. They began to do the math. Were 200 American soldier lives a week worth 1,000 enemy lives taken? Or 5,000, or 10,000? How could even a large number of dead enemy soldiers equal the value of a single American? The answer became a resounding "No!"

This chapter is about the importance of cultural understanding and success in war. This includes understanding our own culture and that of the enemy. If the United States military had defined an outcome in Vietnam that was more consistent with American culture, the citizens may have been more supportive and the military and political out-come of the war may have been different. In twenty-first-century war, where the enemy is defined more by ideology than by politics or geography, failure to understand the nuances of the enemy's cultural worldview can easily lead to stalemate or even defeat.

Winning hearts and minds—both our own and those of the enemy—is no small task. It requires a paradigm shift in the way strategic leaders as well as common sol-diers approach warfare. It also further elevates psychology and allied disciplines to an even more critical level of importance. A serious effort to create a military that under-stands how to leverage cultural factors in achieving its strategic objectives will drive significant changes in military doctrine, tactics, techniques, and procedures.

THIS IS ASKING A LOT FROM OUR MILITARY

Building cultural competence into military doctrine and tactics will not be easy. A good deal of the difficulty lies in how to train soldiers who can be ruthless killers in one instance, and then quickly shift gears and become sensitive and nurturing agents of cultural support the next. This is no small task. In some ways, this requires that the military train its members more like contemporary police are trained. The vast major-ity of day-to-day police work hinges on officers having a good understanding of the community they serve, excellent interpersonal skills, the ability to communicate effec-tively with people, and a genuine concern for the well-being of people they encounter on patrol. However, once in a while, the police officer may have to instantly shift to an aggressive, perhaps even deadly, stance. This continual shifting between social support and mediation to aggression and back again can be extremely challenging to police officers, all of whom have more extensive training in human behavior than do mem-bers of the military.

Traditionally, the military focused almost exclusively on the use or show of force to gain its objectives. From the first day of basic training, soldiers are taught both the warrior mindset and warrior skills. Most people who join the military believe in the sanctity of human life, and it takes a good deal of training and socialization into mili-tary culture to produce a soldier who is not only able, but also willing to kill for his or her country when ordered to do so. Challenging physical training and training in basic combat skills like hand-to-hand fighting and rifle marksmanship accomplish more than honing basic soldier skills, they also toughen the recruit and overcome his or her reluctance to use deadly force on other humans.

This may sound harsh to some readers, but the warrior attitude and willingness to kill is a necessary prerequisite to being an effective soldier, and in forming an effective army. History has taught us that there are times when our nation must defend itself against raw aggression. Negotiation and mediation was not an option following the attack on Pearl Harbor. Osama bin Laden and his associates would likely not have ceased attacks on the United States in the absence of a military response. How many more people would have been killed in the Holocaust if the Allies had not employed their highly trained and deadly forces on the enemy?

But, like their law enforcement counterparts, soldiers who are trained to kill have difficulty turning that switch off and on. In Afghanistan, a platoon may ward off an attack by Al Qaeda insurgents in the morning, help villagers build a sewer system in the afternoon, and negotiate with tribal elders in the evening. It is difficult to tell friend from foe, and soldiers must be on edge and ready to employ deadly force in an instant. Let your guard down and you may be killed. Keep it up, and you may fail to connect with members of the community whose support is instrumental in attaining higher order objectives.

There are some in the military who argue that this is asking too much. It has been suggested that the military identify some units as traditional war fighting units, and others as peacekeeping units. The former would be trained and socialized in the ready use of deadly force, whereas the latter would be trained in skills, like negotiation, that support peacekeeping. It is unlikely this compromise would work. For one thing, in twenty-first-century war there is no clear delineation of battlefield or battle lines. A violent exchange occurs and within minutes or hours life returns to normal. It would not be clear where to position the two types of units. And it is doubtful that the enemy would not employ its own deadly force on the peacekeeping units. Moreover, in an era of declining fiscal resources, having what in essence would be two separate armies would be cost prohibitive.

So the military will be faced with developing cultural knowledge and skills in its forces, along with traditional war fighting skills and attitudes. Its ability to accomplish this challenging task may have a profound influence on our nation's future.

TRAINING CULTURAL PERSPECTIVE

To start, the military must integrate cultural training into every level of its training and education system. For the army, this means cultural training must occur in training new enlisted recruits, and both junior and senior NCOs. For officers, the education of cultural skills must begin during precommissioning training, and continue at all levels of postcommissioning professional military education. It should be a cradle-to-grave program, and endorsed by the highest level of the chain of command as an essential war skill—just as essential as competence in traditional military tasks such as artillery or marksmanship.

Cultural training must not be limited to the military schoolhouse either. Cultural training must be synchronized with other military training to make it a useful tool in actual combat settings. One approach is to integrate cultural-based problems into the curriculum of military combat training centers. The US Army is doing this already, such as at the Joint Readiness Training Center (JRTC) at Fort Polk. Actors are hired to play the part of indigenous people who role-play various dynamic interactions with troops who rotate through JRTC prior to an upcoming combat deployment. Reports from the field suggest this training is very realistic and is perceived as useful.

To be effective, the training must be systematic and the outcomes carefully assessed. Procedures that work should be expanded and those that do not dropped from the curriculum. It must include a wide variety of types of interactions in order to build the mental scripts that soldiers and small unit leaders need in order to perform effectively during deployments.

Psychologists and closely allied disciplines with expertise in cultural and interpersonal behavior should play a significant role in designing, delivering, and assessing this training. The content in both schoolhouse and field training should be based on scientifically valid principles and not simply on the operational experience of military trainers.

In some ways the efforts to integrate cultural training into combat field training exercises is ahead of efforts to include it in formal military education. Just a few years ago, when it became apparent to the senior leadership of the army that the United States was losing face in the eyes of the world due to the many cultural blunders its ground forces were making in Iraq and Afghanistan, it was decreed that cultural training should be significantly increased at West Point. In the end, the answer to this was to double the number of language courses that most cadets take from two to four, and to add additional cultural immersion experiences to cadets such as spending a semester abroad as part of their undergraduate experience.

At the very best, it is not clear how having four semesters of a foreign language builds cultural understanding and the relativistic perspective on values, customs, and religion that are prerequisite in understanding and accepting the way people in other cultures live their lives. Moreover, most cadets enroll in languages, such as Portuguese or French, that are unlikely to be spoken in areas of the world in which we are most likely to fight, or are spoken in disparate countries around the world that share a language but little else.

If exposure to a language, by itself, is not sufficient to build cultural skills, then what is? There may not be a simple answer. In a very large sense, a liberal college education may be the best seedbed for developing this perspective. There may be specific courses in sociology or anthropology that can enhance cultural understanding. It may be that direct experience in a foreign culture could provide, for some people, the crucible event that engenders growth in cultural understanding. For instance, I send one or two cadets each year to participate in a Norwegian Naval Academy field training exercise. These cadets join an otherwise all-Norwegian squad and complete eight to ten days of intense military training. Although the language of instruction is English (the principle language among NATO members), the cadet's intimate interactions with Norwegian peers before, during, and after the training exercise often result in what most cadets say is their most impactful educational experience during their 47 months at West Point.

Psychologists and social scientists must weigh in on the issue of how best to teach cultural skills to the military in the future. Military members cannot learn the language or study the specific culture of every country they may deploy to. Education must be tailored to fit the developmental readiness of rather disparate groups of individuals. The content and methods used to educate young enlisted soldiers, most of whom have completed high school or the equivalent, will not be the same as those used to educate future officers in their precommissioning education, be it through a traditional university, a service academy, or Officer Candidate School. Intermediate officer and NCO education will emphasize tactical and operational implications of cultural understanding. In contrast, cultural education at the level of the senior war colleges must have a strategic perspective, as their graduates will be the future generals and admirals of the military.

BUILDING CULTURAL UNDERSTANDING OF THE ENEMY

Psychologists can help prepare the military by training cultural sensitivity in various ways, as discussed above. But in an unpredictable and volatile world, it is often not possible to predict with any accuracy just where our military may be deployed. The army is currently reorganizing in such a way that certain components will focus on certain regions of the world. For instance, several divisions may be trained primarily to prepare for operations in Asia, others in Africa, and so on. This allows the training to focus on the particular nations and cultures in the designated region. The military can develop a fairly accurate "human terrain map," and inculcate its members with at least some key knowledge of the religion, politics, economics, and cultural practices that are common in the targeted region.

This is a sound approach, but the truth is that it is difficult to predict with any precision where the next war will occur. And with a small military, any utilization of armed force that goes beyond a few airstrikes or very short and limited engagement of ground forces will necessitate using soldiers from all units—regardless of the area of focus for training—in order to project sufficient force to be effective.

Therefore, psychologists and other behavioral and social scientists must also be prepared to assist field commanders during actual combat deployments in learning the cultural topography of their area of operations. This means that these scientists must be deployed in the combat zone with the troops in order to actively research and provide timely information that will help field commanders deal with the cultural component of operations.

This is a radical role for behavioral and social scientists. It is one thing to provide academic instruction in an army schoolhouse to soldiers preparing to deploy to war. It is quite another matter altogether to mobilize these specialists for deployment to combat. It is also worth noting that behavioral and social scientists and their supporting professional organizations, including the American Psychological Association, the American Sociological Association, and the American Anthropological Association, tend to be very liberal if not extremely leftist in political orientation. Finding motivated and properly educated social scientists who are physically able and mentally willing to put themselves into harm's way to assist the military in garnering cultural knowledge of the enemy is no small undertaking. And those who do volunteer may be shunned or stigmatized by other members of their profession.

Human Terrain System (HTS)

In recent years, the US Army has developed just such a program. The Human Terrain System (HTS) program involves forming small teams of behavioral and social scientists to deploy with army units. They are tasked with finding out what the commander needs to know about the local culture, even when (as is often the case) the commander himself does not know what he needs to know![2]

It is worth taking a closer look at the HTS program, as it provides insight into how psychology and related disciplines may influence wars of the twenty-first century. Following September 11, 2001 and the initiation of combat operations first in Afghanistan and later in Iraq, it became clear almost immediately that commanders and their units that were ignorant of local cultural customs and realities were prone to make blunders that could have strategic consequences. Not knowing local cultural practices—at the very least—can make soldiers appear arrogant, insensitive, and rude. At the worst, such behavior may be irrevocably divisive and lead to retaliation against individual soldiers or their units.

I will point out that these issues tend not to present themselves so much during full intensity combat operations. You don't need to be particularly sensitive to cultural nuances when employing deadly force on armed opponents who are trying to kill you. Instead, cultural issues manifest themselves when major combat operations cease and nation-building begins.

One of my colleagues in the Behavioral Sciences and Leadership Department at West Point shared many an instance of the importance of cultural understanding that he encountered while serving as a deputy brigade commander in Iraq in 2004–2005.

As a senior member of the brigade team, he was often called upon to organize meetings with Iraqi citizens and groups of all sorts. These meetings could cover all manner of topics, ranging from settling economic disputes to negotiating solutions to significant political and ideological issues.

My colleague rapidly learned to use cultural practices to manipulate meetings with Iraqi's in order to make certain points. As I described earlier, it is considered rude to openly carry a weapon when invited into someone's home or formal meeting area. If my colleague felt it was important to "make nice" and form a favorable impression, he would turn his M-4 (the rifle most commonly carried by American soldiers) and pistol over to his subordinates, and enter the meeting unarmed. This, coupled with other social graces of the region, enabled him to establish better rapport with his Iraqi counterparts. At other times, when he sensed that it was necessary to make a bolder and harder-edged impression during a meeting, he would arrive in full battle dress, carrying his weapons into the meeting. So in a sense, he would reinforce cooperative Iraqis by treating them with respect. As others became more cooperative, he would change his behavior to be more culturally appropriate. He had many other examples, but this illustrates how knowing simple cultural practices can be used to one's advantage when trying to induce a positive and collaborative relationship between the military and the local people. Contrast this approach with that of many commanders, who failed to show the locals this sort of respect.

The HTS program is designed to provide this sort of information and much more. While the idea that cultural knowledge is important to war is not new, and is reflected at various levels of doctrine, this has been of relatively little value to field commanders. These doctrines "have generally been produced with limited reference to the lived experience of commanders on the ground that actually use this information in day-to-day planning and execution of operations."[3] While doctrine is important to the extent that it formalizes the importance of a given domain (in this case, cultural skills) and suggests generalized ways that the domain must be understood and exploited to enhance mission effectiveness, the complexity and diversity of cultures encountered in modern warfare greatly limits the applicability of doctrine in specific contexts.

In short, what has traditionally been lacking is an empirically based approach to generating useful cultural information for battlefield commanders. The HTS program was designed to do just this. Rather than deriving generalized advice on tactics, techniques, and procedures from doctrine, an HTS team is designed to provide real-time and accurate answers to questions that battlefield commanders raise that allow them to deal more effectively with the local population.

The HTS approach involves forming small teams of four to six members, at least one of whom has a graduate degree in psychology, sociology, or anthropology, and assigning the team to a battalion or higher unit (i.e., brigade or division) that is deployed in combat. The most important skill that the team brings to the commander is the ability to conduct valid field research and to make sense of the data. The research skills common to psychologists—identifying a problem, designing a research plan, data collection and analysis, and sense-making and reporting of the data—are critical to the ability of the team to provide reliable and valid answers to the commander's questions. I would argue that because psychologists generally receive broader and more specialized training in both research methods and quantitative analysis compared with other social scientists, they ordinarily have a better-developed skill set for heading an HTS team. Supporting members of the team may or may not have graduate degrees in a behavioral or social science. At least one should be skilled in the appropriate language. It is important to have a military team member (either active duty or former military with recent operational experience) to allow the team to integrate and communicate effectively with the host unit.

Once formed, the new HTS teams—there were 27 such teams deployed in Afghanistan as of the spring of 2010—undergo a 16-week training program prior to attachment to the assigned unit and subsequent deployment. The content of the training continues to evolve, but includes an intense and in-depth study of the social, cultural, political, and economic structures of the region to which they are assigned. Armed with this generalized understanding of the culture, the team can utilize its research and analytic skills to derive answers to very specific questions a commander may have.

Once deployed, the HTS team is available to help commanders understand the human element of the area of operations. Commanders who appreciate the importance of this understanding—and not all of them do—can task the HTS team to answer specific questions. For example: What local factions exist in the area of operations? Who are the religious leaders and how powerful are they? Often, the commanders may not know what to ask. In this case, HTS teams can formulate their own questions with the goal of providing the commander with knowledge that may become invaluable while conducting operations in the area. It is critical that team-initiated research be operationally relevant. They are not collecting data to test a theory or hypothesis, or to complete a doctoral dissertation. With experience and in conjunction with regular communication and coordination with the military, the team may often uncover important facts about the local culture, even if it is not always clear (even to them) what they are looking for.

Initial reports on the deployment of HTS teams suggest they may indeed provide a useful tool for the commander. In Afghanistan, for instance, some of the research conducted included analyzing the influence of tribal elders, providing an economic profile of certain districts and provinces, and identifying what locals considered to be appropriate gifts. The latter is not inconsequential. Imagine showing up to a party here in the United States and conferring the gift of a goat to the host, instead of a bottle of wine or other culturally relevant gift. Getting off on the right foot matters when soldiers interact with local leaders.

In Iraq, HTS teams provided information on the political and administrative structures of local leadership, identified and analyzed the power basis of different tribes, and analyzed black market trading. Given the Islamic culture, many questions about religion were researched. A recent analysis showed that commanders used HTS teams to learn more about power structure, formal and informal legal systems, economics, cultural beliefs and traditions, and the capability and capacity of local systems as diverse as education, medical and health care, and sewer and water services.[4]

Commanders find this information invaluable in assessing their area of operations. It informs them on who holds the power in the area, and who might be potential allies and threats. Learning local customs and traditions helps the commander interact more positively with the local population. In the nation-building phase of operations, this combines to help the commander establish good relations and rapport with the people. A well-respected and trusted occupying force is likely to receive much more support than one that relies on authority and brute force to effect change.

This is how it is supposed to work. There have been many success stories of HTS teams, but there are also challenges to forming and utilizing effective teams. First, it is hard to recruit motivated and qualified behavioral and social scientists to head these teams. Not many psychologists are willing to leave their families and friends and deploy to a war zone, no matter how fascinating the work may be. There is also a tangible element of danger. In twenty-first-century warfare, simply moving around in the area of operations is dangerous. An improvised explosive device (IED) does not discriminate between an army truck occupied by soldiers and one occupied by HTS members. The enemy frequently assaults military bases with mortars and other indirect fire. So, even as a noncombatant, being in the theater is risky.

In addition, there are other challenges. As mentioned before, there is also pushback from the professional and scientific societies that may view assisting the army as tantamount to engaging in homicide. The team leader needs to be skilled in various research methods, and not just rely on the ubiquitous five-point Likert scale. Someone on the team must be functionally fluent in the local language. Reliance on translators hinders genuine communication and may make data collection far more difficult. Some commanders may not appreciate the value of the HTS team, and may not use it to leverage success in operations.

Despite these possible pitfalls, HTS provides a model for how psychologists may help the military in twenty-first-century war. Even in traditional wars between nations that involve massive engagements between two (or more) heavily armed forces, after the fighting ends the nation-building begins. And to the extent that twenty-first-century war will pit the United States against enemies that are nontraditional and who have significantly different ideologies and religions, psychology and related disciplines will play a major role in success or failure.

Social Network Analysis

A rapidly emerging area of social psychology that is extremely useful in understanding the cultural side of the battlefield is social network analysis (SNA). Human beings behave, communicate, and interact in complex ways. With SNA, it is possible to identify these complex patterns of interaction. Think about an organization that you are familiar with. Let's say you work in an office with 20 other people. It is divided into three subordinate teams, each with its own leader or supervisor. A single manager oversees the work of all three teams. On the face of it, you might expect there to be a predictable and reliable pattern of communication flowing from the manager to and from the teams, as well as between teams. After all, things were organized in this way to achieve this effect. However, if you measure communication among management and teams, and among individual workers, you may find that actual communication involves a much more complex pattern that may differ from what is planned.

In a real organization, the topography of the direction and frequency of information flow may reveal important things about the nature of that organization, such as who the true leaders are. In the hypothetical example above, one of the three team leaders may disproportionately communicate not only with the manager, but also laterally with other leaders, and with individuals of his or her own team and that of others. These interactions can be graphed to reveal a social network, with nodes (representing individuals or, in some cases, subordinate units) with connecting lines that can depict both the direction and frequency of communication, as illustrated in Figure 6.1. A careful analysis of this network can identify who within the organization wields formal versus informal power and who the workers look to for answers and guidance.

Now, consider the case of a commander deployed in Afghanistan. He will probably know who the legitimate authorities are, but he won't likely know the identities of informal leaders and people wielding social power. Social network analysis may provide a tool that psychologists can use to provide this information. Granted, it is more difficult to conduct SNA within the context of a village or town. But there may be data available to generate at least a partial solution. Phone or e-mail records may be secured and examined. Trained observers may monitor the frequency and location of visitors to previously identified local leaders. For instance, maybe the local imam is seen interacting with key locals at three times the daily rate of the tribal chief. In some instances it may be possible to conduct interviews with locals to begin mapping their social networks.

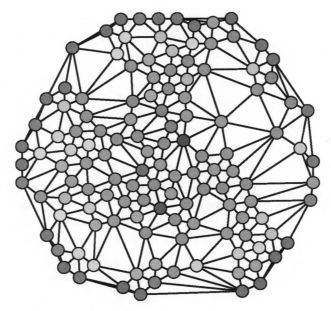

FIGURE 6.1 Example of a social network analysis. Each node represents a different person within an organization. (Image courtesy of Claudio Rocchini.)

The commander will use this information to his tactical advantage. He will know who really exerts formal and informal influence in his area of operations. He will know who to cultivate good relations with and with whom not to waste valuable time. If a crisis occurs—perhaps a soldier or other American is taken hostage—he will know who the deal makers are, and who in the local population influence them. More importantly, he can cultivate good will and peaceful relations by knowing the social topography of his area.

NEGOTIATION TRAINING

Perhaps one of the most important skills needed in twenty-first-century war is the ability to engage in effective negotiations with local inhabitants within the area of combat operations. Clearly this must involve a functional understanding of local culture, history, politics, and economics in order to be effective. Psychologists will play a key role in training the military on how to conduct such negotiations.

An exemplar of negotiation training is the West Point Negotiation Project (WPNP).[5] Started in 2006 by a small group of faculty within the Behavioral Sciences and Leadership Department at West Point, the WPNP has rapidly become a prototype of a negotiation training method that is based on solid psychological principles. Because much of the burden for negotiation in contemporary war falls on small unit leaders (mainly lieutenants and captains) it is critical that very junior officers receive instruction in these skills.

Regardless of whether they are trained or not, military members will by necessity find themselves engaged in negotiation. And it is not limited to members of the local community such as political, religious, or tribal leaders. Negotiations will also occur within the chain of command, with personnel from NATO allies, and even within one's own unit. By sheer luck or brute force, the negotiation may turn out favorably in a given instance, but if the negotiators are not trained in the principles that underlie the practice of negotiation, they will ultimately fail or turn back to military force to gain compliance, something that often causes more long-term problems than short-term gains.

The WPNP uses the results of years of social psychological research as the basis for both a formal academic course in negotiations that may be taken by cadets, and for field training outside the context of the Academy itself for active duty army personnel. The approach is based on a combination of the social psychology of persuasion and compliance, and decision making. For instance, the program suggests a seven-question checklist to guide negotiations:[6]

1. Think about each party's interests.
2. Think about each party's alternatives.
3. Brainstorm solutions.
4. Consider ways to legitimize solutions.
5. Identify commitments each party can make.
6. Analyze the relationships in play and how important they are.
7. Plan your communications strategy.

Cultural understanding is critical to each of these stages, particularly when negotiations occur in the context of significantly different cultures (as in the case of military operations in Afghanistan) and under conditions of high threat. It is one thing to try to talk a car dealer into a lower price for a new car; it is quite another to negotiate the release of a hostage in a combat environment.

Cultural competence is fundamental to negotiations. Without it, steps 1 and 2 are virtually impossible. Knowledge gleaned from the HTS team will help, but military negotiators must be able to look at things from other cultural perspectives. The solutions generated in step 3 and planned in step 4 must be culturally acceptable. Steps 5 and 6 hinge on an understanding of the local power structure and knowing who has social power within a community.

Step 7, planning the communication strategy, relates to a rich history of research on persuasion that identifies three components to persuasion: the communicator, the message, and the audience.[7] In the context of war, the communicator is likely to be a platoon leader or perhaps a company commander. According to persuasion theory, the communicator must be viewed as credible. Credibility, in turn, requires the communicator to be viewed as both competent and trustworthy. The platoon leader must quickly establish himself, in the eyes of the local community, as someone who knows what he is doing. Trust may be even more important. The platoon leader who does not follow up on promises or who treats people inconsistently and rudely is not likely to be an effective negotiator.

A particularly effective method for learning negotiation skills is through role-playing and reinforced practice. Toward that end, the negotiations course at West Point includes a block of instruction that provides cadets the chance to practice what they have learned in the classroom in a realistic setting similar to what they are likely to encounter following commissioning when deployed in combat zones. Actors are hired to play the part of tribal leaders, religious figures, and political heads. The actors are given a script to follow and cadets are tasked with negotiating—across several different stations—to obtain compliance with their requests. These negotiations are closely monitored by expert observers, and cadets receive immediate and detailed feedback concerning their performance. The field training occurs in conjunction with a negotiations conference, which includes guest speakers and workshops conducted by recognized negotiation experts. Systematically designed, and based on sound psychological science, this negotiation training may provide a model for similar training at military training centers and major commands. Figure 6.2 shows cadets engaged in a negotiation exercise.

The message included in the negotiation is also important, as indicated in the steps outlined above. Years of research suggest that, when trying to persuade someone to comply with a given course of action, it is important that the message convey both (or all) sides of alternative courses of action, and that the content of the message is subject to refute. This is often a hard lesson for military personnel to learn. It is easier to fall back to a simple one-sided order than to objectively and fairly present multiple sides of an argument. But research shows this is especially important, particularly where initial opposition to an idea is high, as often may be the case in combat areas.

Finally, understanding the audience is critical to negotiations. The commander who understands, through his own training and experience or through information from HTS teams, the local culture is much more likely to get his way. Social psychologists inform us that people and cultures may differ in their need for factual, logical arguments. They differentiate between "central route" persuasion, which is heavily based on the compelling logic in the message conveyed to the audience, and "peripheral route" persuasion, which depends more on other factors such as the emotional appeal of the message.[8] Add to that the nuances of local custom, and effective negotiation skills are clearly something that do not develop by chance.

FIGURE 6.2 Cadets negotiate with Army Major Stephen Flanagan, in the role of local police chief, during a negotiation exercise at West Point. (Photo courtesy of Neil Hollenbeck.)

The social psychology of compliance is also relevant to negotiation in military contexts. Social psychologists have identified and researched various methods used to get people to comply with requests.[9] To comply with a request does not imply, necessarily, that a person initially agrees with the request or wants to do it. The following examples of compliance techniques described by social psychologists are presented with a short discussion of how they apply to military negotiations.

Norm of Reciprocity

This simply means that when you do something good for others, they are more likely to reciprocate by doing something good for you in turn. A commander who builds a new sewer system or facilitates the building of a school or health clinic earns important capital and good will that may be returned by compliance to some future request.

Door-in-the-Face Technique

This approach involves asking the target to do something they would never initially agree to, but quickly follow up with a lesser, more reasonable request. Any number of examples come to mind, but in using this technique the commander knows that his initial request is likely to be rejected outright, but he then stands to look good when he offers a reasonable compromise.

Foot-in-the-Door Technique

In this approach, the commander initially makes small, more easily accepted requests, and gradually increases his demands once compliance to the initial demand is met. At first, for instance, he may ask key members of the local community to simply attend a weekly meeting to discuss issues and community needs. Later, he may ask individuals to commit greater time and resources by serving in leadership positions or giving time to specific projects. In any case, with this technique, the locals may not have responded to the more intrusive requests to begin with, but having acquiesced to lesser requests, are more willing to take on larger ones later.

Lowballing

Here, a commander asks a tribal leader to commit to some action but then, before the initial action is completed, increases the demands or costs of that same action. This is a common technique that sales personnel frequently inflict upon us. When buying a car, we may negotiate a price with a sales associate and sign an initial agreement—only to learn after the sales representative speaks with his or her manager that the actual cost of the car is much higher. A clever commander can use this, along with other techniques discussed, to improve compliance with the policies he implements in his area of operation.

These applications of basic social psychology methods to military settings seem very simplistic to anyone with formal education in psychology. But military personnel do not know these techniques, and to be effective in twenty-first-century war they must learn them, and learn how to employ them with skill. Sales and marketing personnel have certainly used these techniques on you many times, and you may resent being manipulated in this way. Indeed, while manipulative in nature, it is far better to negotiate with the enemy and their supporters than to kill them, something that twenty-first-century war will try to minimize. Psychology, therefore, plays a significant role in equipping the military with these cultural and social skills to facilitate effective relations with members of the indigenous populations where our military will find itself operating in future wars.

UNDERSTANDING THE TERRORIST

Psychologists must help the military understand terrorists and other nontraditional foes in twenty-first-century war.[10] Although there are hundreds of articles published

on terrorism and terrorists, only 100 or so scholarly articles on the psychology of terrorism have been published. The majority of these are either conceptual or employ relatively weak research designs such as interviews with incarcerated terrorists. Conceptual articles attempt to extract from the tenants of a given psychological theory the characteristics of a terrorist. In defense of those who study terrorists and terrorism, well-controlled and methodologically sound research protocols are virtually impossible to carry out in studying terrorists. As a rule, terrorists go about their business under the cloak of secrecy, and are not particularly given to granting interviews with researchers or completing surveys prior to executing their terrorist acts.

Nevertheless, providing insights into the nature of the terrorist may help the military develop strategies to identify and intercept potential terrorists before they accomplish their mission, or, in a more general way, develop operational programs and procedures that diminish the motivation of individuals or groups to engage in terror acts.

Several approaches to this problem currently exist. Some researchers focus on understanding the motivation of individual terrorists. For instance, some terrorists may behave in this way because of an underlying psychopathology that drives them to engage in homicidal behavior. According to this view, the individual who engages in terrorism may suffer from some biological or psychological imbalance. For instance, various personality disorders, including narcissistic and borderline personalities, have been speculated to contribute to terrorist behavior.

Other psychologists look for patterns of personality traits that describe terrorists. The aim is to identify personality profiles of terrorists. In addition to personality, demographics can also be profiled. It is important to note that the demographics of who is a terrorist may be subject to cultural stereotypes, and also may change with time. The prototypical terrorist of the late 1960s may be quite different from the prototypical terrorist at the current point in the twenty-first century. Also, it may be that rather disparate types of individuals are drawn to different forms of terrorism. An eco-terrorist may not resemble (in terms of age, ethnic or religious background, education, etc.) a terrorist inspired by religious jihad.

Moreover, the predictive power of the personality trait approach is weak at best. Even if statistically reliable personality correlates of terrorism are identified, they may lack sufficient validity to serve as a practical and useful predictor of who is or may become a terrorist.

Most personality trait approaches focus on maladaptive traits that correlate with some sort of psychopathology. The emerging field of positive psychology, in contrast, has identified 24 positive character strengths thought to be universal in humans.[11] I have found in my own research that some of these traits, including honesty, persistence, and self-regulation, are stronger in military members than in the general population.[12]

Some of these traits correlate with leadership. Interestingly, from this point of view, terrorists may score relatively high on certain positive character strengths. There is no empirical evidence to suggest what positive character traits terrorists would likely possess, but one might speculate that many would score high in some of the very same traits that predict leadership and other adaptive behaviors in other venues, including members of our own military. While this may seem paradoxical—how could a terrorist have positive character strengths?—it should be remembered that one culture's terrorist may well represent another culture's freedom fighter. Terrorism, it seems, may be in the eye of the beholder.

Some psychologists look at terrorism more from a social-psychological lens, focusing on the impact of socialization and group dynamics as etiological factors in the development of a terrorist. According to this view, becoming a terrorist is more than faulty individual development. Rather, it is a social process heavily influenced by culture, social norms, and socialization.

There are many other psychological interpretations of terrorism. To engage in terrorism may be a rational choice, one based on a highly valued goal of improving the state of affairs of one's religious, social, or national group. It may be a learned behavior, something that leaders and role models may socially condition in their followers. Durkheim's taxonomy of suicide may provide a description of the motivation behind at least some suicide bombers.[13]

Military psychologists are well positioned to empirically investigate terrorism and terrorists. Forward deployed psychologists, like those in HTS teams, may be able to systematically interview and observe people who identify as terrorists or who express strong sympathy for terror as a sociopolitical weapon. However established, an empirically based study of terror and terrorism will be a significant contribution made by psychologists in wars of this century.

CULTURE MATTERS

The ideas discussed in this chapter represent just a few of the important ways that a psychological perspective on culture may influence the training and education of the military, and in assisting battlefield commanders in their day-to-day operations of understanding the population they are dealing with in general, and with potential terrorists in particular.

This is an area where psychologists may gain much by partnering with their colleagues in related behavioral and social sciences. Linguists, anthropologists, and sociologists—along with psychologists—will have a major voice in the planning and execution of wars of the twenty-first century. The nations that best utilize the talents of these scientists will have an edge in fighting future wars.

NOTES

1. Kristi Keck, "U.S. Must Win Afghan Hearts and Minds, Commander Says," *CNN. com*, Sept 28, 2009, www.cnn.com/2009/POLITICS/09/28/afghanistan.obama/index. html?iref=allsearch.
2. Montgomery McFate, Britt Damon, and Robert Holliday, "What Do Commanders Really Want to Know?" 92–113.
3. McFate et al., "What Do Commanders Really Want to Know?" 93.
4. McFate et al., "What Do Commanders Really Want to Know?" 108.
5. For more information about the West Point Negotiation Project refer to the project website at www.usma.edu/wpnp/SitePages/Home.aspx.
6. Jeff Weiss and Jonathan Hughes, "Implementing Strategies in Extreme Negotiations."
7. A good discussion of the social psychology of persuasion can be found in any introductory psychology text. See for example Michael W. Passer and Ronald E. Smith, *Psychology*, 631–32.
8. Richard E. Petty et al., "To Think or Not to Think: Exploring Two Routes to Persuasion," 81–116.
9. These techniques and related research are described in all introductory psychology texts. Again, see Passer and Smith, *Psychology: The Science of Mind and Behavior* (see note 7).
10. See Ragnhild B. Lygre and Jarle Eid, "In Search of Psychological Explanations of Terror," 114–28.
11. Christopher Peterson and Martin E. P. Seligman, *Character Strengths and Virtues.*
12. Michael D. Matthews et al., "Character Strengths and Virtues of Developing Military Leaders," S57–S68.
13. Emile Durkheim, *Suicide.*

7

THIS IS NOT YOUR FATHER'S ARMY

This is a democratic country and the military must represent the country. And then externally, with just the expansion of missions, the places we're going, the challenges that we have, the demands that we have—culturally, ethnically, mission-wise, wherever we go or where we're going to go in the future, diversity is an absolute must for us.
Admiral Michael G. Mullen, *USN (Ret.)*[1]

My father used to say that "the army joined me" in 1943, which was his way of saying he was drafted. He was a bit older than many draftees at 29 years of age, but in other respects was representative of Americans—both draftees and volunteers—who coalesced to form the largest military in US history. He was white, had a high school diploma, and came from a rural background. Born in Tea, Missouri, he had never traveled outside the state at the time he was drafted. He had little to no exposure to diverse cultural or racial groups, nor any experience with people from other nationalities. His lack of social sophistication didn't matter much. World War II was a kinetic war. He only had to learn to fight. His draft letter said he was to serve in the US Army for the duration of the war plus six months. The harder that he and his fellow soldiers, airmen, sailors, and marines fought, the sooner he would come home. He and millions of others did just that.

The military of World War II was overwhelmingly white and male, especially at the beginning of the war. African Americans were largely excluded from the officer corps, and frequently draft boards, usually comprised completely by white males, did not select African Americans for military service. Those who were drafted or volunteered served in segregated units, usually commanded by white officers. Similarly, women were represented in only small numbers in World War II. They were not subject to the draft, could not serve as career officers or NCOs, and were severely limited to service in support roles. In short, the World War II military did not represent the demographics of the United States. One might speculate that Japanese soldiers may have concluded that the United States was entirely a white nation.

Much has changed since the end of World War II. Formal barriers that excluded service by African Americans and other minorities are gone. Women may now serve in

all military occupational specialties, including those that involve direct combat. Gays and lesbians can now serve openly without fear of being dismissed from the service. Members of all racial and ethnic groups serve in the military, which is arguably the most diverse employer and social institution in the country.

From a purely operational point of view, the race or gender of a soldier does not matter so much in a traditional kinetic war. Sheer military force is more important. It doesn't matter a whole lot if the person who drops a nuclear weapon on an enemy target is white, African American, Hispanic, male, or female. The target will be destroyed, and when enough kinetic energy has been released onto the right targets, the war will be won. Both world wars of the early and mid-twentieth century were these types of wars. This is not to say that psychological and sociological factors did not play a role, but they were less critical in determining the outcome of those wars than they are to contemporary war and, presumably, future wars.

Diversity matters to the military for reasons beyond social and occupational equity and opportunity. As we have seen in Chapter 6, cultural understanding plays a crucial role in twenty-first-century war. The kinetic phase of the Iraq war lasted only a few weeks. President George W. Bush was technically correct when he announced on May 1, 2003 that major combat operations had ceased. But his proclamation revealed a fundamental misunderstanding of the nature of current war. Defeating the Iraqi army did not equate to winning the peace. The senior political and military leadership of the United States made a series of blunders that laid the groundwork for years of violent civil unrest and terrorist attacks that ultimately took the lives of thousands of American military personnel, and government and contract civilian employees. Many of these mistakes resulted from sheer ignorance of Iraqi cultural terrain or ethnocentric arrogance.

Employing cultural understanding in military operations is facilitated if the military itself is culturally and socially diverse. Such a military will have, particularly within the ranks of a large military like our own, personnel who originate from the culture in which military operations are occurring. They may speak the language and understand the customs and history of the country or region in which military operations are taking place. They may be invaluable in helping commanders plan strategies to win over the occupied population, if the military learns how to utilize this storehouse of knowledge and expertise. The US military faces global threats. It is hard to predict where the next war will occur. It could be the Middle East, North Africa, the Korean Peninsula, or in Central or South America. Recruiting and maintaining a diverse military, therefore, is critical to national defense.

As poorly as the military has responded to cultural aspects of operations in the recent past, imagine how much less effective it would have been if the World War II military fought in Afghanistan or Iraq. After the kinetic phase of operations, male, white, Christian, rural soldiers would suffer greatly in trying to communicate effectively and win over indigenous populations. Their homogeneity would further alienate them from the people, and make negotiation and forming common goals more difficult.

The concept of diversity in the military will continue to develop along lines that are somewhat different from traditional representation across racial, ethnic, and gender lines. The changing nature of warfare will allow other types of people who in the past have been perceived as unfit for military service to fill important jobs within the ranks of the military. With advances in prosthetics, amputees and other severely wounded soldiers will be able to continue to serve. With technical knowledge and expertise becoming more important to the operation of weapons and other military systems than mere physical strength, restrictions on weight, eyesight, and other physical characteristics may be lifted. Obesity, in and of itself, should have zero impact on the ability of a person to operate an unmanned aerial vehicle (UAV). The future military will be open to an increasingly diverse array of people, and this diversity should strengthen the military and make it more effective in dealing with all manner of threats.

THE UNITED STATES IN 2050

The racial/ethnic composition of the United States will continue to change dramatically through 2050. At the beginning of the twentieth century, only one of every eight Americans was a race other than white. By the end of the century, one of four was nonwhite. The Pew foundation projects that by 2050, non-Hispanic whites will account for 47 percent of the population.[2] Much of this increase will come from rapid increases in the Latino population. By 2050, they will comprise 29 percent of the population, compared to 14 percent in 2005.

Moreover, and of significance to the military from the standpoint of cultural diversity, in 2050 nearly 20 percent of Americans will be immigrants. These immigrants will come from a variety of countries, and possess cultural knowledge that will be invaluable to the military. Because the military has traditionally been a means for immigrants to learn important job skills, have a secure income, and build better prospects for their own children, it is reasonable to project that first and second generation immigrants will step forward disproportionally (compared to other groups) to serve in the military. Psychologists and other social scientists can help the military learn to leverage this diversity of culture and perspectives to improve the "cultural IQ" of the military, as it continues to deploy in diverse cultures around the globe.

Americans in 2050 will also be older. Pew projects that the number of elderly will double by 2050. The number of children and working age Americans will continue to shrink. Life expectancy for a man in 1945 was 63.6 years, and for a woman was

67.9 years. By 2000, these numbers were 74.3 and 79.7, respectively. The MacArthur Foundation Research Network projects that in 2050 the life expectancy for US males will range from 83 to 86, and that of females from 89 to 94.[3]

The US military has rather severe restraints on the ages of people it will employ on active duty. Generally, one must be 35 years of age or younger to join the military, and most are much younger than that. With the exception of general/flag-rank officers and very senior warrant officers and NCOs, military members face mandatory retirement after a maximum of 30 years of service. For someone who enlists in the military at age 18, this means they will be forced to leave at the relatively young age of 48. Given the life expectancy projections for 2050, this means that if the current policies are in effect then, a recent military retiree may live another 40 years or more. This has all manner of implications, including the tax burden resulting from paying pensions for such a long period of time. These relatively early retirements also deprive the military of an abundant pool of experience and expertise. Since many military jobs do not truly require the physical strength and stamina more common to young soldiers, it may be to the military's advantage to change their policies and allow members to serve longer. If this happens, then psychologists may play an instrumental role in preparing both the military as an institution and individual military members to prepare for remaining on active duty longer in life, perhaps into their sixties. This will also affect the social dynamics of the military in ways that may be difficult to project.

Extrapolating from these trends, it is likely that the US military in 2050 will be more racially and ethnically diverse, older, and, with the few remaining restrictions on the service of women now lifted, more female. From an organizational psychology perspective, this implies changes in social power and group dynamics. At the end of World War II, a soldier was most likely white, male, young, and from a rural background. In 2050, the soldier will be more likely to be nonwhite, perhaps equally likely to be male or female, and from an urban background. With large numbers of recent immigrants in its ranks, the military can grow cultural competence from its own members. A diverse military will be necessary to maintain effectiveness in the sorts of missions likely to typify twenty-first-century operations.

CURRENT RESEARCH AND FUTURE PROJECTIONS

Psychologists and other behavioral and social scientists play a vital role in understanding and building diversity in social institutions including the military. We have the conceptual and analytic skills to assess diversity, identify areas of concern, and recommend ways of increasing diversity. Ironically, the military has both a long history of discrimination on the basis of sex, race, and sexual orientation, and at the same time an excellent track record of remedying these problems once it is tasked to do so. Over the next few pages, I will review where the military currently stands with respect to diversity efforts for various groups of people, discuss current research that identifies where improvements are needed, and project the impact of future demographics on efforts to maintain a diverse and varied military force.

African Americans

Currently, African Americans make up 22 percent of the US Army overall. This is similar to their representation in the other services. This represents an overrepresentation when compared to the general population of the United States, where African Americans make up less than 13 percent of recruiting-age men and women. Also consistent across the services, African Americans comprise much lower percentages of the officer corps. For example, the percentage of African American officers in the army is 12.3 percent. And, for the combat arms branches (e.g., infantry, artillery), fewer than 10 percent of officers are African American. Combat support jobs that have high transferability to civilian occupations, such as transportation and administrative jobs, have attracted higher proportions of African Americans—both officers and enlisted—during the era of the all-volunteer force.[4]

In many ways the integration and utilization of African Americans in the military is a success story, especially when viewed in the historical context of racism and prejudice that were rampant in the military until fairly recent times. The true integration of African Americans did not begin until after World War II, and race riots plagued the military as recently as during the Vietnam War. With the advent of the all-volunteer force in the late 1970s, however, African Americans (and other economically or politically stigmatized groups) have seen the military as a stepping-stone on which to improve their lives. As we have seen, African Americans have responded by joining the military in large numbers. Unlike other major social institutions, the military does not discriminate as a function of race or gender on important economic factors like wages. A master sergeant earns the same pay, regardless of race, religion, gender, or ethnicity.

Colonel Irving Smith, an associate professor of sociology at the US Military Academy, argues that despite these very real successes in the utilization of African Americans in the US Army, too many African American officers still fail in comparison to their white counterparts in the army.[5] He maintains there are many reasons for this. For example, African American officers make up only 6.7 percent of the generals in the army. The causes for this are subtle and varied. One factor is, as mentioned above, that African American officers tend to select support and administrative jobs in the army, possibly because of high transferability of skills later to the civilian economy. However, the warrior culture of the army reinforces officers who select traditional combat arms—such as infantry—with high promotion rates relative to support jobs.

It follows that the lower representation of African Americans in combat jobs results in lower promotion rates to general officer. This is not a case of blatant discrimination, but rather an insidious effect of a culture of economic disadvantage that propels many African Americans to select army jobs with higher civilian transfer at the expense of high probabilities for promotion to the rank of general officer.

It is also possible that the underrepresentation of African Americans at the general officer rank is a cohort effect. That is, currently serving general officers entered active duty over 30 years ago. Economic, political, and social forces have changed over those 30 years. Colonel Smith also reports that the percentages of African Americans in the various officer ranks, from second lieutenant through general officer, have increased dramatically compared to their numbers in 1997. With the exception of first lieutenants (where African Americans decreased by 6.7 percent), all other ranks have shown substantial increases. Notably, the number of African American full colonels has increased 144.9 percent. This is particularly significant because it is from the ranks of colonel that the next generation of general officers is selected.

One solution to increase the representation of African American officers at the higher ranks is to improve recruiting for both the service academies and for ROTC programs. Because the mission of the service academies is to educate and train the future senior leaders of the military, the greatest gains may be from increasing the numbers of African American candidates at each academy. West Point, for example, has long maintained certain class composition goals that include forming a corps of cadets that at a minimum reflects the percent of African Americans graduating from college.

While West Point has been successful in meeting class composition goals for other minorities, it has routinely struggled to reach the goal with African Americans. In recent years, the percentage of African American cadets has been in the vicinity of 7 to 8 percent, well short of the 12 percent class composition goal. As early as 1968 systematic efforts were initiated to recruit qualified African American candidates for admission to West Point.[6] Although West Point admitted its first African American cadet in 1869, there were never more than 10 African American cadets in any given class until 1968. In 1960, West Point opened an equal opportunity admissions office, followed in 1976 by Project Outreach, each aimed at aggressively increasing African Americans in the corps of cadets. By the early 1970s, the number of African American cadets admitted increased several fold. Today, about 100 African American cadets accept appointment each year to West Point.[7]

Despite these successes, West Point and the other service academies continue to strive to recruit additional African Americans to reach a number that is representative of their number in the general population, or about 12 percent. For West Point, that would mean admitting about 50 additional African Americans each year. One obstacle to meeting this goal is the high admissions standards that West Point and the other service academies have. All candidates must score very high in academics, leadership potential, and physical fitness. All of the service academies compete with elite private and public institutions. Because the pool of qualified African Americans is smaller, the competition to recruit African Americans for the academies is fierce. Elite universities can offer scholarship packages that rival those of the service academies, and without incurring the burden of a lengthy obligation for active duty military service.

Psychologists may aid the military in developing recruiting strategies that make service academies appear more attractive to African Americans. Besides appealing to the patriotic value of serving in the military, perhaps recruiting efforts can also highlight the practical benefits that attending a service academy have for its graduates after they complete their initial obligation of military service. Service academy graduates are leaders in all domains of public service and private enterprise in the United States, and their alumni-support organizations are among the strongest, most unified, and most effective of any institutions of higher education in the country. Educating possible recruits about the long-term benefits—especially in terms of building future connections to aid in personal and career growth—could enhance the perceived value of a service academy education compared to that of a traditional college or university. In short, psychologists may be able to help the services frame a more effective approach to recruiting African Americans for admission to their academies. Coupled with parallel efforts at ROTC programs, over time the services could significantly increase the numbers of qualified African American officer candidates, who would go on to serve in the highest ranks of the military in the future.

Women

The past few decades have seen a major shift on how women are utilized in the military, as well as how their suitability for military service is viewed by the general population. It would have been unthinkable in World War II for women to serve in military jobs that exposed them to the dangers of combat. Today all military jobs are open to women. Women were first admitted to the military academies in 1976, with the first classes at the Naval Academy, Air Force Academy, and West Point graduating in 1980. They have served with honor and distinction in Iraq and Afghanistan, to include exposure to direct combat.

In a move that would have been unthinkable to many just a generation ago, in January of 2013 the Secretary of Defense, Leon Panetta, rescinded the ban of women serving in direct combat jobs in all branches of the US military. Given the nature of modern war, the exclusion of women from combat jobs did not make a lot of sense. There are seldom front lines, and regardless of their job, all military personnel deployed in a combat theater are vulnerable to attack. Moreover, for all practical purposes, women have served in combat for many years.

The opening of women to all military jobs has secondary impacts that are very important. This will shatter the glass ceiling preventing women from attaining higher rank more so than any other policy change in the history of the US military. Because combat leaders are more likely to be promoted to general officer, over time, as women serve in these jobs, they too will obtain the necessary prerequisite experience needed to assume high rank. This should also increase the ability of the armed forces to attract and retain the most competent women, both in the officer and NCO ranks, who previously were deterred from serving in the military due to institutional barriers to success.

The indications are that the military will make the selection to combat jobs based on objective criteria that are gender neutral. For example, to be an infantry soldier a man or woman would need to meet certain physical characteristics, such as the ability to carry a fully loaded combat pack, negotiate obstacles, and operate certain weapons systems. Instead of assuming that all men and no women can do these tasks, soldiers can be assessed during basic training before being selected for additional infantry-specific training. Indeed, the army is already developing new and fairer physical fitness standards that may be used to select and screen all soldiers—both men and women—for entry into combat jobs.

Such a policy change may meet with some resistance in both the military and in the general society, at least at first. With respect to resistance within the military, my first experience as a behavioral sciences officer in the air force may shed some light on what today's policy makers may be up against. I had just arrived at my first duty assignment shortly after receiving my commission as a second lieutenant. My unit was tasked with conducting basic and applied research on personnel and human resources. An order had come down the chain of command from Air Force Headquarters for psychologists in my unit to do an empirical study of the impact of increasing the number of women in the air force on operational readiness. The very title of the tasking revealed a bias that, at the very best, the only impact of such a course of action would be negative, and a desirable outcome would be no negative impact.

The order trickled down to the lieutenant colonel who was my direct supervisor. He had a doctor of philosophy in experimental psychology and had taught in the Behavioral Sciences and Leadership Department at the Air Force Academy. He formed a team of seven psychologists, including me, and we developed a research plan. First, we developed both a survey and an oral interview protocol designed to be given to worker level air force personnel (typically junior airmen and NCOs), along with parallel instruments given to their immediate supervisors. Our intent was to assess the work experiences of air force personnel with respect to various dimensions of tasks they could typically engage in, and then ask their actual supervisors questions that focused on task performance and whether gender played any role in that performance.

In all, we surveyed several hundred air force personnel, both workers and supervisors, at locations within the continental United States and in the Pacific and Europe. We gathered data on over a hundred different tasks and dimensions, and carefully analyzed the results for evidence of gender-related differences in performance.

In the process of collecting the data, we interviewed the base commander at each installation we visited. In almost every case, these commanders offered up anecdotal "evidence" of why women were not effective in jobs that required physical strength, such as firefighter. Remember that this was 1980, and women were just beginning to make inroads in nontraditional jobs like police officer and firefighter in the civilian sector. There were many opinions on their suitability for these jobs, but not much fact. Amazingly, almost every commander told us the story of a plane that had crashed on landing and caught fire. The various versions of the story had one common theme. Unfortunately for the male pilot, the responding firefighters included a woman who was unable to pull the pilot to safety before the aircraft burst into flames. If a man had been there, he would have had the strength to save the pilot's life. Case closed: women should not be in certain jobs!

Of course, it seemed odd to us that the story was so similar. We began checking the facts of the case and found that we could not verify that such an incident had ever actually occurred. I do not suggest the commanders were intentionally lying, but I do think they were exemplifying the innovative and productive nature of human memory, blending together elements of real events with hardened beliefs, to unconsciously invent a story that supported the notion of barring women from certain jobs. Moreover, simple logic tends to argue against the veracity of this story. It is hard to imagine an all-women firefighter crew in that era. When it comes to pulling a pilot out of a burning airplane, everyone would have been fighting for the chance to help. I doubt they would send just one person—woman or man—to do such an important job.

This brings me to the outcome of the study and more importantly our experiences in briefing it to the air force senior chain of command. First, the results. Of all the comparisons we made, we failed to identify any jobs or tasks that women could not perform as well as men. There were a couple of tasks—such as lifting a heavy toolbox from the ground up to the wing of an aircraft—which somewhat more women than men reported difficulty with. But they quickly developed work-around solutions. In the case of the heavy toolbox, they simply asked a coworker to help them lift it—as did smaller men. Aircraft maintenance, like virtually every other job in the military, is a group task. There are always other people around to lend a hand.

What was most interesting was the reaction we got when we briefed our findings. We traveled to the Pentagon and briefed a two-star general. My lieutenant colonel supervisor gave the briefing. I had the easy part—in the time-honored tradition of a military briefing, someone had to flip the transparency charts (this was long before PowerPoint). The general was seated at the front of a room, with an entourage of other senior officers (colonels and one-star generals) behind him. It was an intimidating set-up. To make matters worse, my boss had broken his ankle while running the week before, and had to give the briefing while seated, with his leg—in a big cast—propped up in front of him. I don't think the general liked this at all.

The briefing consisted of approximately 35 transparencies. It started off fairly well, and as we covered the background of the study and the method, the general nodded and hurried us along. When we got to the results, we presented slide after slide showing no adverse impact of women on mission readiness or job performance. With each passing slide, the general seemed more emotional. Finally, his face turned beet red and he yelled "STOP!" at the top of his lungs. He removed his glasses, wiped sweat from his face, and proceeded to inform us (still at the top of his lungs) that he did not need "a bunch of high-fallutin' PhDs to come and tell me a bunch of stuff that ain't true!" In the interest of prudence, I have omitted the more colorful words that he used to emphasize his point. But the message was clear. The headquarters brass had expected us to find evidence that increasing the number of women into the air force would significantly and negatively impact its mission to fly and fight. After his outburst, my boss coolly said, "Sir, would you like me to complete the briefing?" The general muttered a few more curses, but allowed as how he had wasted this much time on the briefing, he might as well hear the rest. We went home, thinking our careers were over, but fortunately they weren't. I hope things have changed a good deal in the intervening 30-plus years, but I suspect there continue to be a number of very sexist if not blatantly misogynist individuals in the military service.

One other finding of this study is worth mentioning. While we found that increasing the number of women in the air force would have no adverse impact on mission readiness, we did find that increasing numbers of dual military couples with children and single parent members did adversely impact mission readiness because of the difficulty of assigning dual members to the same base, and having realistic plans for child care when one or both parents deployed. A lot of strides have been made since 1980 with respect to policy on this matter and structural support for such members, such as enhanced child care facilities on military installations. Single parents and (with more women entering the armed forces) dual military couples are more common today than ever before. Hence, to maintain combat readiness, the Department of Defense must continue to improve support programs so these military members are available to deploy on short notice.

Negative stereotypes with respect to the role of women in the military also exist in the general culture, despite years of progress in breaking down barriers to women that exist in the workplace. I have conducted research on the attitude of Americans toward women in the military. Utilizing a set of questions from the General Social Surveys, I have looked at whether people believe a woman should or should not serve the military in the following nine military roles: jet fighter pilot, truck mechanic, nurse in a combat zone, typist in the Pentagon, military commander, hand-to-hand combat soldier, jet transport pilot, air defense gunner, and crew member on a combat ship.[8] In general, the strongest support is for women serving in noncombat jobs (e.g., truck mechanic) or in jobs that women traditionally occupy in the general society (e.g., nurse, typist). The weakest support comes for two types of jobs: first, jobs that involve direct combat (especially hand-to-hand combat soldier), and, interestingly, military commander. Men are less approving than women. And traditional college students (that is, those enrolled at private and public universities without an explicit military affiliation) are more approving than either service academy cadets or ROTC cadets. Importantly, this biased attitude seems uniquely American. A sample of Norwegians indicated almost 100 percent approval of women serving in comparable jobs in their military.[9]

I also have found little evidence that service academy cadets have become any more approving over the past 25 years, at least at the beginning of their service academy tenure. I first surveyed Air Force Academy cadets on these nine jobs in 1988.[10] Over 25 years later, I asked the same questions of West Point cadets. The pattern of approval rates was stunningly similar, usually within plus or minus one percentage point. In each case, I surveyed cadets in their first of four years of study at their respective academies. At West Point, we now have data on cadets in their first-class year (seniors). We have not published these results yet, but it is clear that after four years the attitudes of both male and female cadets become much more broadly accepting of women serving in diverse military roles. Education seems to have worked.

In many ways, the story of the integration of women into the military parallels that of the integration of African Americans. Tremendous strides have been made, but there is a lot of work left to do. Even with access to most or all military job specialties, women all too often face challenges ranging from subtle paternalistic behavior to felonious sexual assault. Widely publicized sexual assault and harassment scandals have occurred at all of the service academies, where despite over 30 years of gender integration, a strong masculine and macho culture persists. The military remains one of the few occupations where the number of men greatly outnumbers the number of women. Part of the solution may be empowering women to serve fully by lifting any remaining quotas on the number that may wish to do so. For example, West Point currently limits the percentage of women in each new class of cadets to approximately 15 percent, a number linked to the representation of women in the army at large. It is interesting that class composition goals for African Americans are based on the percentage of African

Americans in the recruiting age population. If this was the standard for women, then the class composition goal would rise to a more equitable 50 percent. At West Point, and likely throughout the military, achieving a balance of men and women more typical of general society may help address many of the current problems that women experience while serving in the military.

Psychologists can take findings from the military and from general society and use them to help educate and train its members in diversity. Social psychologists may suggest training scenarios that can be used in enlisted basic training and precommissioning courses to break down gender stereotypes, and help both men and women members learn to serve without bias and prejudice.

Gays and Lesbians

For over 30 years I have studied the issue of how disenfranchised groups of people—chiefly women and minorities—have been viewed with respect to suitability for military service. I wasn't around for the beginning efforts to integrate the military along racial lines, and I began my career as a military psychologist just as women were beginning to make significant inroads into the military. So it is with some sense of excitement that I am present to experience the full integration of gays and lesbians into the military.

Until the 1990s, gays and lesbians were explicitly prohibited from serving in the military. If a soldier, sailor, airman, or marine was discovered to be or admitted to being homosexual, they were immediately discharged. Moreover, the mere suspicion that a military member was homosexual would justify an investigation into his or her status, and if indeed the subject of the investigation was found to be homosexual, they would be discharged. Under President Clinton, the so-called Don't Ask, Don't Tell (DADT) policy was initiated. Under DADT, as long as a service member quietly went about being homosexual—in essence lived a secretive life in terms of their sexual orientation—they could continue to serve the military. This constituted the "don't tell" part of the policy. Under the "don't ask" part, military commanders could not ask a member questions about their sexual orientation. Despite this policy, there were widespread reports of homosexual military personnel being investigated and discharged from the service because of their sexual orientation.

After the election of Barack Obama as President, the Department of Defense moved to repeal DADT and to allow gays and lesbians (and other nontraditional gender/sexual identities) to serve openly and with full status in the military. Several military research and policy organizations, including the Department of Behavioral Sciences and Leadership at West Point, were tasked to investigate the various ramifications that such a change in policy would have for the military. Ultimately, the repeal came late in 2011. The new policy is elegantly simple. Sexual orientation has no bearing on the recruitment, hiring, assignment, promotion, and retention of military personnel. Gay and lesbian organizations are now recognized on military installations in the same way as other social organizations. Homosexual couples may live openly.

I almost added that homosexual couples may live openly without fear of reprisal. At the time of this writing, less than a year following the repeal of DADT, I have been somewhat astonished by the lack of rumor or publicized instances of discrimination, hostility, or outright aggression against military members. I suspect such incidents do occur, but I am encouraged that there has been so little controversy within the military on this change.

To say this surprises me is an understatement. Just a few years ago, Aaron Belkin, a noted expert on gays and lesbians in the military, spoke to a group of faculty at West Point. He gave an excellent historical perspective on how homosexuals have been treated in the US military, and presented a cogent argument in favor of the repeal of DADT. While most faculty listened with open minds, several stood and vigorously voiced their objections, mostly on religious grounds. One lieutenant colonel waved a bible (he had come prepared, it seems), cited Leviticus, condemned all homosexuals to hell and perdition, and said he and most of the officer corps would resign if and when gays and lesbians ever served openly in the military. He was not alone in his reaction. It was interesting to hear an argument in favor of gays and lesbians serving openly in the military, based on reason, empirical evidence, and the constitution, contrasted with one grounded in somewhat peculiar interpretations of religious dogma. I imagined members of the Taliban giving similar arguments against homosexuals serving in the military.

I will share a few more anecdotes from West Point on this subject before reviewing a bit of research. One observation was that when DADT was repealed, one female army major at West Point reportedly informed her immediate chain of command that she was a lesbian. If she had wanted a significant reaction, she must have been disappointed. Her colleagues evidently suspected that she was a lesbian anyway, and were more impressed by her excellent officership and job performance than they were by her sexual orientation.

The anecdote captures cadet reactions to the repeal of DADT too. Compared to the current generation of cadets, I grew up in a completely different social environment. In my high school graduating class of about 300 students, homosexuality was never discussed and no students openly admitted to being homosexual. In hindsight, we suspected a couple of classmates of being homosexual, but we didn't talk about it. It was sort of the DADT of its time.

In contrast, these days I ask my cadets each term if they knew any high school classmates who were homosexual. Everyone raises their hands. I then ask how they reacted to having openly homosexual classmates. This is where they look at me like

I am a fossil, and inform me that it is not a big deal to them. When DADT went away in the fall of 2011, I asked them how it would change the Corps of Cadets. Basically, I got a response of "Gee, sir, we already know who the gay and lesbian cadets are—this isn't a big deal!"

It seems that much of the pushback to the repeal of DADT came from older generation officers and other decision makers. I graduated from high school in 1971. The senior leadership of the military—the three- and four-star generals, the politically appointed civilian leadership of the Department of Defense, and other senior leaders in the military—were socialized much like I was, coming from the same generation. It is fair to point out, however, that it was precisely this age cohort who exhibited the leadership to change the policy.

I have surveyed various groups of individuals regarding the jobs they believe that homosexuals should—or should not—fill in the military. Focusing on the same nine military jobs that I described earlier in my studies of the attitudes toward women serving in the military, I have asked service academy cadets, civilian college students, and students from Norway whether a gay male or a lesbian should be able to serve in each of the nine jobs.[11] Interestingly, for US samples, the results parallel those of approval of women. Specifically, there is widespread support for gays or lesbians serving in noncombat roles, and fairly substantial pushback for serving in either direct combat jobs or in positions of command. Among the Norwegians, where homosexuals have openly served in the military for some time, there is almost no objection to them serving in any military role.

Military psychologists must help the military track and evaluate the progress on its integration of gays and lesbians. We can help the military develop strategies to facilitate this process. Much of the classic social psychological research on competition, cooperation, and prejudice is relevant. The wars in Afghanistan and Iraq have in some ways become true-life manifestations of Muzafer Sherif's classic study of competition and cooperation among boys in camp.[12] In this era of more open sexuality, heterosexual and homosexual soldiers have fought together against a common enemy and emerged with a better understanding of themselves and greater acceptance of others who may be different in terms of sexuality. The repeal of DADT allows all service members to personally know openly gay and lesbian colleagues. The mere exposure effect may quickly erode existing prejudices. As one colleague put it, "Homosexuality is easier to dislike at a distance . . . when you know the individuals personally, their sexual orientation becomes virtually irrelevant to other, more important aspects of character."

Overcoming Stigma

There are similarities in the objections raised to African Americans, women, and homosexuals serving in the military. History tells us that African Americans were believed to be mentally incapable of most military duties, and certainly not of sufficient intelligence to serve as officers. Women were viewed as sufficiently intelligent, but were considered emotionally and physically unprepared for the rigors of military jobs and leadership. Finally, gays and lesbians were seen as intelligent and perhaps emotionally and physically fit, but their character was questioned. All three groups were stigmatized, but for different perceived deficiencies.

Through policy changes brought on by changes in the social, political, and economic climate of the country, the military has discarded discriminatory policies. We have seen, in the case of African Americans and women, that it takes time for full integration to occur. The military should learn from this lesson, and aggressively employ all the tools at its disposal to hasten the full integration of homosexuals, as well as to completely dissolve remaining obstacles to the full integration and utilization of African Americans, other minorities, and women. Psychologists will continue to play a crucial role in this process through the next 25 years.

Other Stigmatized Groups

Physical fitness has long been a signature trait for military personnel. For infantry soldiers and personnel specialists alike, the military has a one-size-fits-all standard for physical fitness. The culture of fitness goes beyond pragmatic considerations of possessing the physical skills and strength to complete mission requirements. It has become, in the modern military, an engrained part of the culture. Military personnel with the ill fortune of having bodies that do not match the stereotype of a lean, muscular soldier may find themselves at a disadvantage for promotions and key assignments, and if they fail to meet stringent height and weight standards, may be expelled from the military.

In many military occupations these physical fitness standards make sense. An infantryman, who must carry extremely heavy loads on his (or soon, her) back for long distances, negotiate difficult obstacles, and perform physically draining tasks for hours or days at a time must be in the peak of physical condition.

It is hard to overstate the fervor with which the military embraces the physical fitness culture. In an environment where no cadet would ever think of making jokes about a person's race, gender, or sexual orientation, they do not hesitate to openly express disdain and disgust for people even mildly overweight. It is well known in the army that several of our recent four-star generals selected their aides at least partly on the basis of them being able to keep up with them on daily runs. In meeting a new comrade or commander for the first time, soldiers will instantly make a negative evaluation if that person is overweight, regardless of knowledge of how competent they may otherwise be.

With the supersizing of American society as a whole, it is increasingly difficult to recruit soldiers who can attain the physical fitness and weight standards of today's military. It is likely that these standards will change to be more inclusive of different shapes, sizes, and physical abilities. UAV pilots need to have good eye-hand coordination, excellent working memory capacity, and well-developed decision-making skills. But it does not matter if their body mass index (BMI, a metric of the optimal ratio of height to weight; anything over 25 is considered overweight, over 30 obese) is a slim 19 or a rotund 35.

Another major area of change in the military that is likely to continue well into the twenty-first century is the ability to retain amputees and other soldiers with serious physical wounds, who until recently would have been forced into retirement. New technologies in artificial limbs coupled with the emerging science of brain machine interface will enable more and more soldiers to serve despite their injuries. If wounded soldiers can serve with these constraints, then it may be possible to open initial enlistment or training for service as an officer to people with significant physical disabilities.

We already need a new word to describe people like the modern amputee. A generation ago, the terms "disabled" or "physically challenged" may have been appropriate. In the 2012 Olympic Games, Oscar Pistorius, a double amputee, represented his country of South Africa in the 400 meter track and field event (both the individual and team relay events). It is hardly fair to call this person disabled. Maybe a better term is "reenabled": reenabled through technology, and ready and able to serve in the defense of their country.

Psychologists are critical to these efforts on a number of fronts. Counseling and clinical psychologists help the affected soldier come to terms with the injury. Biological psychologists are helping design artificial limbs that interface directly with the central nervous system.[13] Research conducted by psychologists with primates has demonstrated that monkeys can learn to operate artificial limbs via self-generated cortical changes. If we can learn to control artificial limbs through cortical activity, it should be possible to wire the sensory system of the artificial limb to the brain, resulting in a limb that not only is functional in terms of motor output (walking, grasping, etc.), but also feels just like the real limb to the individual.

Summary

The military of 2050 will be less white, more gender balanced, open to all individuals regardless of sexual orientation or gender expression, and include all body types and more people who have successfully overcome significant physical injuries. I am certain that the average soldier of World War II would be astonished at the changing face of the twenty-first-century military. Psychologists will play a pivotal role in making these transitions successful.

UNDERSTANDING GENERATIONS X, Y, AND Z

There is a common perception that newer generations of Americans are somehow different from preceding ones. Young men and women of recruiting age today are sometimes thought to have short attention spans, different values, less commitment to long-term goals, and a different worldview than older generations. If this is true, then psychologists can help identify ways that this and following generations differ. This will guide recruiting, training, and career development—in short, all aspects of military personnel operations.

Whether there are substantive differences between these new generations and older generations is an empirical question, and one that psychologists and other social scientists are examining. How do emerging technologies affect how and what we learn? Does nearly universal access to computers and other keyboard devices change the way we communicate via written language? Do teens really communicate more via text messages than in face-to-face interactions? If so, does this have an effect on their ability to work together in small social and work groups that comprise the backbone of the military?

It is fair to suggest restraint in making assumptions about what young people know, don't know, do, and don't do. For example, there is a widely held belief that all teens are very computer savvy. The US Army did an empirical study of the computer skills of new recruits, and found that most did not have a working understanding of the most basic and common computer tasks.[14] Because the army is digitizing so many of its systems, if it had simply assumed that new soldiers would be adept at using them, this would have adversely impacted training and operational effectiveness.

My West Point colleague, military sociologist Morten Ender, and I have been looking at the so-called millennial generation, especially with respect to how they view military-relevant issues and questions. Specifically, we are comparing the attitudes of service academy cadets, ROTC cadets, and traditional college students with no formal military affiliation on topics including attitudes toward military service, views on war, and beliefs about diversity in the armed forces. Today's millennials will be tomorrow's strategic leaders in the military, government, and business world. As such, their attitudes about these issues will influence how the nation construes and employs its military. We have found, for instance, that support for the wars in Iraq and Afghanistan was influenced heavily by gender and political affiliation,[15] and that gender was also a significant factor in support of expanded roles for women in the military.[16]

THE HUMAN DIMENSION: FORMAL DOCTRINE FOR A CHANGING ARMY

I will close this chapter by observing that the army, at least, is very much aware of the vital importance of understanding its own soldiers, their families, and army civilians in developing its future forces. The army is in the process of developing new doctrine that formally defines the human dimension of warfare. This emerging doctrine will guide how future soldiers are trained, educated, and led. Although this particular doctrine is specifically for the army, the other services are also recognizing the importance of a formal understanding of human behavior and are working on similar additions to their doctrine.

The Army Human Dimension (AHD) doctrine focuses on understanding the soldier of 2020.[17] It anticipates uncertainty in missions that may include foreign (all levels of potential conflict, from peace support operations to full-scale war) and domestic (for example, assisting in natural disasters) operations. It also includes army civilians, who will play an even more vital role to the success in the army in coming decades, and army families.

A quick look at the attributes of the soldier of 2020 underscores the role that psychologists will play in developing this future force. That soldier should be of high character, physically and emotionally fit, self-disciplined, willing and able to learn new tasks, adaptable, and intelligent. Moreover, this soldier should be able to be a team player, culturally savvy, technically proficient, and be able to learn new tasks easily. The future soldier—even at the lower enlisted ranks—will be expected to be diplomatic. And, of course, the army of 2020 will continue to be more diverse.

The future soldier will perform in very trying circumstances. It is unlikely that the nation will wage war with a peer foe. The major wars of the twentieth century are not likely to be duplicated. While potential threats will include some regular armed forces (e.g., the Iraqi army in 2003), the majority are more likely to be irregular, terrorist, and even criminal in nature. Ideological groups and some countries will continue to resent the US presence in their region. Although they will not have the firepower to confront and defeat US forces directly, they will engage in tactics designed to erode the will of the nation to continue the fight.

Continued innovations in technology will place additional demands on soldiers. As we have discussed before, it won't be enough for a soldier to be physically strong. The soldier of the future must be able to understand, use, and repair complicated digital systems. Robotics, drones, new generation weapons—all of these tools will utilize the latest computer technology to improve lethality.

The AHD aims to employ evidence-based behavioral and social science to improve its ability to recruit, assess, train, develop, manage, and retain soldiers and army civilians. The doctrine formally employs three lenses to focus on these tasks: psychological, physical, and social. You can consider these lenses as something of a Venn diagram. Thus, to understand adaptability, scientists must inform policy makers and trainers on the cognitive components of adaptive behavior, the neurobiological factors involved, and social components to adaptive behavior. For any given behavior, one lens may provide more useful information than another, but in almost every case it is necessary to consider the contributions of all three to develop a useable understanding of the target behavior.

The psychological lens includes cognitive and emotional processes, motivation, self-control, and mental toughness. Psychologists can bring together current research and theory to aid in developing the psychological skills of the future soldier, but also may engage in new research to help identify and better understand other primarily psychological factors that affect soldier performance and adaptation. The physical lens includes brain science, nutrition, physical fitness, hormone function, and similar primarily biological processes. Finally, the social lens focuses on things like cohesion, esprit de corps, family and social relations, and cultural understanding.

It is almost a cultural revolution for the army to elevate psychology to this level of importance. In doing so, it is worth pointing out that the AHD will guide future funding in psychology and related sciences. This will range from very basic research on, for example, the neurobiology of stress and resilience, to applied research on building cohesive teams. By necessity, the army (and the other services) will turn to academic scientists at universities across the nation to help it build the scientific basis of AHD. As history tells us, this research may yield findings that cause quantum advances in psychological science, and hopefully will help prepare a strong and adaptive future military.

SUMMARY

Two historically significant changes are shaping the military of the future, and psychologists find themselves at the center of both. First, social, technical, and demographic changes are producing a military that would be unrecognizable to combatants of twentieth-century wars. Psychological expertise will guide and inform how these changes affect recruiting, training, and operations in the future.

The second change—and from my perspective this is truly revolutionary—is the formal emphasis that the military is placing on psychological science. Even 25 years ago, the idea that the military would view psychology as a difference maker in its operational success would not have been seen by the military itself as viable. But changes in the nature of war, technology, and world politics have caused psychology to emerge as the science that will make the difference in wars of the twenty-first century. In the end, war is simply a means of attaining the political goals of a nation. The goal of twenty-first-century war is not to kill or be killed, but to manage military force to win over the support of the nations and people who threaten our strategic interests.

NOTES

1. Admiral Michael G. Mullen, US Navy (Ret.), speaking at the Naval Sea Systems Command Diversity Summit, Naval Surface Warfare Center, Carderock Division, Sept. 18, 2007; see www.navy.mil/search/print.asp?story_id=32208&VIRIN=51370&imagetype=1

2. See "How America Has Changed Over Last 100 Years: Census Analysis Shows Great Changes in US Demographics" at http://usgovinfo.about.com/library/weekly/aa122102a.htm#race.

3. The MacArthur Foundation Research Network on an Aging Society, www.agingsocietynetwork.org/node/156.

4. For a detailed review of the history of blacks and other minorities in the military, see Karin De Angelis and David R. Segal, "Minorities in the Military," 325–43.

5. Irving Smith III, *Why Black Officers Still Fail*.

6. Jeffery K. Toomer, *A Corps of Many Colors*.

7. Smith, *Why Black Officers Still Fail*.

8. Michael D. Matthews, Eddie C. Melton, and Charles N. Weaver, "Attitudes Toward Women's Roles in the Military as a Function of Gender and Race," 426–30.

9. Michael D. Matthews and Jarle Eid, "The Role of Women in the Military: An International Comparison."

10. Michael D. Matthews, "Women in the Military," 212–16.

11. Michael D. Matthews, "Comparison of US and Norwegian Cadets and College Students on Approval of Homosexuals Serving in the Military."

12. This fascinating study is described in Muzafer Sherif et al, *The Robbers Cave Experiment*.

13. For a fascinating review of how the brain may control motor function, see Richard A. Andersen, Eun Jung Hwang, and Grant H. Mulliken, "Cognitive Neural Prosthetics," 169–90.

14. Jean L. Dyer and Geoffrey H. Martin, *The Computer Background of Infantrymen*.

15. David E. Rohall, Morten G. Ender, and Michael D. Matthews, "The Role of Military Affiliation, Gender, and Political Ideology on Attitudes Toward the Wars in Afghanistan and Iraq," 59–77.

16. Michael D. Matthews et al., "Role of Group Affiliation and Gender on Attitudes Toward Women in the Military," 241–51.

17. The Army Human Dimension doctrine is currently being revised. My comments here are based on this preliminary development. For updated information, refer to the US Army Training and Doctrine Command website at www.tradoc.army.mil/#.

8

WHEN THE GOING GETS ROUGH, THE ROUGH GET GOING: LEADING IN COMBAT

Never forget that no military leader has ever become great without audacity. If the military leader is filled with high ambition and if he pursues his aims with audacity and strength of will, he will reach them in spite of all obstacles.
Karl von Clausewitz[1]

Imagine that you are leading 30 soldiers from the LOD (line of departure) to the objective (a gun emplacement near a lake). The terrain is mountainous and, while there are trails, they are dangerous to traverse due to land mines and the possibility of ambush. Your infantry platoon consists of highly trained individual soldiers and small teams equipped with special skills and weapons. Your team is likely to have a machine gun, several fully automatic rifles, and soldiers with expertise in grenades or other explosives. Besides mines and enemy soldiers who will give their lives to protect the target, you also may be threatened by enemy aircraft or artillery.

Did I mention that you only have three hours to accomplish the mission? The success of the larger mission, already in progress, hinges on you taking out the gun on time. If you are late, soldiers in adjacent units will be killed by this gun. That said, this is not a suicide mission. You must try to complete your mission with no deaths or injuries to your soldiers. Along the way, your commander wants timely and accurate updates on the progress of your mission. You want to be viewed as brave and decisive by your commander, and therefore you feel a great deal of additional pressure to succeed.

And what prepared you for this mission? You are a second lieutenant who completed a bachelor's degree just over a year ago. Luckily, you attended ROTC and had courses on leadership theory plus some field leadership training. You completed the Infantry Officer Basic Course prior to being deployed. Your senior NCOs have seven or more years of military experience, and all have been in combat before.

By comparison, imagine that you have been asked to lead a group of 30 young adults on a hike in the mountains, let's say from your point of origin (a trail head) to a camping spot by a lake. Assuming you have a basic understanding of map reading (or, more likely these days, your GPS batteries are fully charged), you could readily organize your hikers and lead them to the lake. The threats you face might include someone twisting an ankle, getting lost temporarily, or being annoyed by insect bites. Nobody is trying to kill you. I expect that people would vary somewhat in their ability to lead in this context, but anyone with enough training and practice could likely get the job done. You can imagine other routine leadership scenarios in business or other settings.

It is a reasonable argument that leading in these two different scenarios are not the same, although some of the basic skills, such as map reading, are similar. Nevertheless, they require different psychological, social, and organizational skills. Before leading 30 young adults on a hike, you might stop and think about the "theory X, theory Y" theory of leadership. I doubt such a thought would cross the platoon leader's mind, however briefly.

One of the immutable truths of war is that leadership matters. This is especially true under the conditions of combat which has been captured in the widely used military acronym (another acronym, of course), VUCA. This stands for *volatile, uncertain, complex,* and *ambiguous.* While many corporate settings can be described as VUCA, most people do not equate corporate leadership—however stressful or important in monetary terms—to combat leadership. To capture the challenges of combat leadership, you must add the very real possibility of imminent death or serious injury for you or those around you, and the political repercussions of taking, or failing to take, certain courses of action. In combat decisions must be made very quickly, and there is no luxury of systematically comparing different approaches to the problem. In the end a leader must select a course of action, and, through the combined force of his or her position and strength of will, compel others to move into the face of deadly fire. And do so in a skilled and orderly manner.

As a research psychologist, I have always found it difficult to conceive of leadership from a scientific perspective. If you ask 10 leadership scholars, they will probably each give you a different definition of the term. In scientific language, it is hard to operationalize the concept of leadership. Yet when you experience good leadership, you know it, even if you can't nail down just what it is that makes that leader good. My squadron commander in Officer Training School was such a person. I didn't find him especially personable, but for reasons I still can't articulate, we wanted to do what he told us. When he spoke, we listened. When he directed, we complied. Most importantly, we did so because we *wanted* to.

It is this ineffable component of leadership that makes it challenging from a psychological perspective. What individual behaviors comprise leadership? Can we understand leadership simply through applying what social psychologists know—and they know a lot—about the dynamics of compliance, control, obedience, and persuasion? Cognitive psychologists can identify the unique problem solving, decision making, and communication skills that effective leaders possess. Organizational psychologists can address the impact of formal and informal roles and how they play out in leading others.

Any organizational psychologist will tell you there is no dearth of theories about leadership. You could spend months or years reading and studying the various theories. Most are rather interesting and help us understand different aspects of leadership. Some are based on empirical observations, but most are based on corporate experience or mere speculation. A common thread is that most theories of leadership accept the premise that leading in combat is not substantively different from leading in corporate settings. In fact, much of our collective understanding of leadership is based on a model of extrapolating from studies of managers and leaders in corporate settings and assuming that the factors predictive of effective leadership in those settings apply equally well in combat.

For the military, perhaps the most important question is whether combat leadership is fundamentally different from leadership in other domains. Do good leadership skills in garrison predict or translate into effective combat leadership skills? Should we take our best combat leaders and make them the strategic leaders of the military? There are no firm answers to these questions, but they are important ones to answer if we are to improve leadership—both combat and noncombat—in the military and civilian organizations.

In the past five years a science of combat leadership has emerged. Inspired by the thoughts of Brigadier General Thomas Kolditz, US Army (Ret.), who introduced the idea of in extremis leadership, a growing number of psychologists have begun to unravel the components of what makes a good combat leader.[2] This represents a paradigm shift in leadership science, because it is now becoming possible to identify what makes a good combat leader, and reverse engineer what makes a good leader or manager in a noncombat setting.

In the remainder of this chapter, I will describe the concept of combat leadership, what makes this type of leadership (and, hence, leader) unique, and speculate on how psychology can contribute to a better understanding of leading others in combat. Over the coming years psychology can help the military refine its understanding of combat leadership, which will have tremendous implications for training and leader development. Much of this may be of value in other extreme, if not mortal, leadership settings.

LEADING IN EXTREMIS

Who would you rather follow into combat: Captain Park, whose main motive for being in the army is pay and benefits; who learns only what the army tells him he has to know in order to get promoted; who directs others into combat but avoids exposing himself to fire; who shares little with his soldiers in terms of hobbies, interests, and tastes; and whose behavior, while technically competent, does not inspire trust? Or Captain O'Neill, who loves his job and his soldiers; who strives to learn everything he can about his job; who "leads from the front" by working side by side with his soldiers, despite the risk; who shares values and interests with his soldiers; and who is viewed as competent and trustworthy? This choice is, to use a colloquialism, a no-brainer.

Tom Kolditz, in his groundbreaking book on the topic, described his own research that focuses on people who lead in situations where lives are on the line.[3] Using a variety of methods, he studied combat leaders, elite parachute team members, special forces soldiers, leaders of mountain climbing expeditions, and others who are responsible for leading and directing the actions of others in situations that include the potential for death or serious bodily injury. From hundreds of interviews and surveys, he began to find a consistent pattern of traits, skills, and attitudes that these in extremis leaders shared. These included:

- Inherent motivation for the task
- Embracing continuous learning
- Sharing risk with their followers
- Having a common lifestyle with their followers
- Having and inspiring high competence, trust, and loyalty

Kolditz is not suggesting that all military members, police, or fire personnel share these five traits. Remember, he only studied in extremis leaders who were successful, chose to remain in their occupation, and in many cases were the best of the best in their field. That said, it is very interesting to examine traits that such leaders share.

The first two traits—inherent motivation for the task and embracing continuous learning—are easily seen in effective combat leaders. Being a platoon leader, company commander, or battalion commander is not a nine-to-five job. Training for combat, leading and guiding your unit during the combat deployment, and recovering from combat are all-consuming tasks that require a complete immersion into the task. Leaders who are inherently motivated for the task and who find intrinsic rewards in preparing their subordinates for the mission devote every ounce of their energy to the task.

Positive psychologists give some insight into such leaders. They suggest there are three types of happiness in life. The least sophisticated is hedonic happiness. This is happiness that derives from feeling good, such as eating a candy bar, enjoying an alcoholic beverage, or perhaps a warm bath. The next level is engagement, which refers to the ability to become absorbed in a task. This is often accompanied by losing the sense of time passing and experienced as a sense of flow. The highest level of happiness relates to the meaningful life. Meaning is related to doing something greater than just for oneself. Transcending selfish motives, the meaningful life results in a sense of having made the world better for those around you. Interestingly, the meaningful life may not always include a great deal of hedonic pleasure. Doing things that benefit others can be painful and can require lots of self-deprivation.[4]

Effective in extremis leaders can thus be described as living the meaningful life. The company commander believes that his actions have a direct effect on the well-being of his soldiers, other units, and the nation as a whole. Such leaders can put aside the physical and psychological strain of a combat deployment and view that as simply a condition to be dealt with that is outweighed by the greater good of effectively leading his subordinates.

It can be especially damaging to such leaders if circumstances arise that cause them to question the meaning of their mission. Military psychologist Paul Bartone has written on the existential angst among military leaders that resulted from the discovery of wide spread prisoner abuse at Abu Ghraib. Realizing that members of your own organization are engaged in activities that run counter to the very core beliefs of the organization can have devastating effects on morale, Bartone believes.[5]

Inherent motivation is closely linked to a commitment to continuous learning. Good military leaders strive to be experts in every domain of their craft. The effective infantry officer never stops studying tactics, strategy, and military history. Learning never stops, because the task of leading others in conditions of mortal peril requires that the leader know as much as he or she can about the job.

Shared risk means that the leader is willing to expose him or herself to the same threats that other members of the team face. This does not mean they do the same job—the leader most often must perform supervisory or directory functions during in extremis situations, but they must be perceived by followers as being "in the fight." While this aspect of effective in extremis leadership may seem obvious to the reader, not all leaders have the ability or willingness to expose themselves to danger and at the same time make competent decisions and other leader actions. This may in fact be one of the most defining differences between in extremis leaders and those who are effective in other situations, but not when under mortal threat. Thus, one could be an excellent commander of a training organization but fail when placed into the field. An engaging example of this can be seen in the HBO miniseries *Band of Brothers*. Lieutenant Sobel, responsible for the initial training of Easy Company, 2nd Battalion, 506th Parachute Infantry Regiment, in the 101st Airborne Division, was a very capable—if not especially well-liked—military trainer. However, he proved incompetent at leading troops in combat, and was relegated to staff work for the remainder of the war.

In extremis leaders who share a common lifestyle with their followers tend to be admired and respected, and inspire confidence on the part of followers. This does not mean that the leader associates frequently with followers in informal settings on a peer-to-peer basis. The leader and followers, especially in the military and paramilitary organizations like law enforcement, may lead somewhat separate personal lives. But they live in similar circumstances, share core values, have similar interests, and most importantly share a passion and pride for their jobs. There is often a clear path for followers to become leaders. Enlisted soldiers can grow into NCOs, or complete college and become officers. Police patrol officers can be promoted to leadership positions. The differences between the leader and the led are ones of degree, not of kind.

The last traits identified by Kolditz are competence, trust, and loyalty. Noted leadership scholar Patrick Sweeney conducted research on these traits during actual combat operations in Iraq in 2003. Prior to the start of the war, he was a doctoral student in social psychology at the University of North Carolina, preparing for a follow-on assignment at West Point in the Behavioral Sciences and Leadership Department. As the army prepared for war, David Petraeus (a major general at that time, and commander of the division) asked Sweeney to suspend his doctoral studies and assume the job of liaison officer and deep fires and effects coordinator for the army's V Corps. He readily agreed, and capitalized on this opportunity to study leadership in actual combat—something that has rarely been accomplished in the history of leadership studies.

Sweeney found that above all else, highly regarded combat leaders—platoon leaders, platoon sergeants, and company commanders—had to be viewed as competent by their soldiers if they were to be viewed as good in extremis leaders. However, Sweeney found competence, while necessary for effective leadership, was not sufficient in and of itself. Good leaders had to have the trust of their subordinates. No matter how competent a leader was, if his soldiers did not trust him to genuinely care for their best interests, he would fail as an in extremis leader. Finally, effective leaders were loyal to their subordinates. They certainly enforced standards and sometimes made difficult decisions, but the soldiers sensed that these leaders would stand up for their best interests and support them in the toughest times. Those leaders that possessed all three of these qualities were highly regarded and effective. But those who were lacking in one or more of these critical traits were not viewed favorably by their soldiers. Having the full confidence and support of subordinates is uniquely critical when leading others in situations where death or physical injury is distinctly possible.[6]

How do these highly competent in extremis leaders do in routine leadership or management situations? There is little empirical evidence to address this question. For the military, one can argue that there is no such thing as a routine, nonthreatening situation. Even in times of peace, combat units must maintain readiness and this means training under realistic conditions with the understanding that the unit could be sent to war with little advanced notice. Ask any army officer who has led a unit in war games at the National Training Center if they felt that this experience was routine.

One can think of anecdotes that address this question. Could a leader like George Patton, who by all accounts was a highly effective—if somewhat brash—field commander during World War II, be an effective chief of staff of a peacetime army? One might speculate that some of the very traits that made him effective as a leader in combat would interfere with competence in a more managerial position like chief of staff. But this is only speculation since I do not know of any empirical studies that have looked at this issue.

It is also true that the military tends to promote to its highest ranks those who are perceived as being its best warriors. In the air force, not only is it is a distinct advantage to be a rated pilot when it comes to promotion to general officer, it is even better to be a fighter pilot. And better still if you have flown and led successfully in combat. In the army, combat arms (infantry, armor, artillery, etc.) officers have an edge in promotion to general officer over those—often highly effective—officers who specialized in combat support positions. Thus, the culture of the military supports the idea that combat leaders make the best overall leaders, but whether this is really the case is largely unknown.

BUILDING IN EXTREMIS LEADERS

It is critical for the military to be able to develop in extremis leaders. Pat Sweeney and I, along with Paul Lester, recently edited a book addressing this. We identify three distinct but related components to in extremis leadership. These are (1) enhancing one's psychological body armor, (2) learning to influence others when people are in harm's way, and (3) leveraging organizational structure and resources to maximize leader and follower performance under these conditions.[7]

Enhancing Psychological Body Armor

Military leaders must possess certain individual skills and characteristics in order to be successful when leading in dangerous contexts. Some of these may be trait-like and difficult to develop, but many others can be learned. Psychology will play a significant role in the future in forming a better understanding of these characteristics and skills and developing strategies for learning them.

Courage

Courage is a fundamental aspect of leading in dangerous contexts. By courage, we mean to include not just physical bravery, but also the moral courage necessary to make hard decisions that carry life changing consequences. What is courage?

Paul Lester and Cynthia Pury maintain that elements of courage include self-volition, a worthy goal, and significant personal risk.[8] A soldier may perform an act that is dangerous without knowing of the danger ahead of time. Most people would consider this action as less courageous compared to the soldier who realizes the danger in advance of the decision to act, and performs the act anyway. Exposing one's self to threat of death or serious injury for less than a worthy cause may be considered foolhardy or rash, but if the act addresses a highly valued end state, then we may interpret the actions as courageous or even heroic. And there must be significant personal risk in order to attribute the behavior to courage. Riding a roller coaster may be frightening for some people but there is virtually no risk of injury. A soldier who is defusing a roadside bomb, in contrast, will be viewed by almost everyone as possessing courage.

Military training and culture encourages the development of courage, both physical and moral. Leaders are encouraged to chose the "hard right over the easy wrong" when making decisions for themselves and their subordinates. During training, honing physical fitness skills and progressively exposing individuals to ever more challenging tasks may desensitize the soldier's fear and increase the propensity to respond effectively in the face of real danger. Organizational climate and peer influence are intentionally developed to create a culture of courage. Exceptionally courageous individuals are awarded medals for their acts. These heroes are venerated by the military and presented to its members as ideals of the virtue of courage.

That said, developing individual courage is far more an art at this point than a science. Future military psychologists must further deconstruct the concept of courage and develop new ways of training and developing it. Courage consists of behaviors, attitudes, beliefs, and emotions. At least some of these may presumably be identified and trained.

Once learned, the leader may take certain actions to foster courage among subordinates. Lester and Pury suggest several ways to do this. The leader should first and foremost serve as a role model, and promote learning through observation. When subordinates see a respected leader behaving in a courageous manner, they may desire to do the same. The leader must also practice his leadership skills and train combat relevant skills among subordinates. Developing competency with the tools of the trade—be they rifle marksmanship or expertise in flying high-performance aircraft—is thought to provide the confidence needed for subordinates to behave courageously when needed. The leader can directly reinforce skilled performance, encourage mental practice, and promote vicarious learning. Coaching in the form of social persuasion and positive feedback may be provided. Reinforcing the ideals of the organization such as selfless service, integrity, and courage may help instill and maintain conditions that set the occasion for courageous behavior.

Stress Management

Stress management also plays a key role. In this case I am referring to combat leaders teaching their subordinates how to best understand and deal adaptively with stress. I will not discuss the definition of stress—you can find a very good explanation of stress in any introductory psychology book.[9] But it is important to understand that stress results from the interaction between an individual and the environment. Because individuals vary a great deal in experience, education, psychological health, attitudes, expectations, and the like, people may react in quite varied ways to similar situations. In combat stress some may react in a pathologic manner and develop depression or other maladaptive responses. Others may be resilient and show little change. Still others may derive positive growth from their combat experience.

Psychological science suggests a variety of techniques and practices that military leaders might employ to help their subordinates respond more favorably to combat. Army Colonel James Ness recommends several tactics.[10] Leaders should first educate their soldiers about the nature of stress. Soldiers read a lot about PTSD, but hear very little about what a normal response to combat stress may entail. It is important they understand that taking a life or seeing others killed or wounded may almost always be followed, for a time, by disturbed sleep and intrusive thoughts and dreams. They must understand that humans are wired to respond in this way to help prepare us to avoid similar situations in the future.

Next Ness recommends that units train continuously. This speaks to the notion of building competence. I would suggest the term "hypercompetence" to describe a degree of performance that ensures almost complete automaticity of combat skill tasks. As a police officer, I did not fear making felony car stops or disarming suspects because I had been trained how to do it. Overlearning critical skills thus reduces the perceived stress inherent in certain situations.

For Ness, unit cohesion is also a central part of a prescription for effectively dealing with combat stress. Highly cohesive units such as Army Rangers and Special Forces have the lowest rates of PTSD and other pathologic responses to combat stress. Military discipline and teamwork combine to protect soldiers against stress.

Ness also recommends that leaders establish a culture of catharsis, wherein soldiers feel comfortable sharing their reactions to experiences they have encountered. Finally, leaders and the military as an institution must work tirelessly to develop ways of teaching effective coping strategies to soldiers. Absence of pathology is not a sufficient goal. Future military psychologists must find evidence-supported methods of teaching soldiers how to cope with the fear and stress that is inherent in combat. A good starting point may be the concept of hardiness, which includes the components of challenge, commitment, and control. As discussed in Chapter 2, hardiness is trainable and soldiers who are high in hardiness respond more positively to combat stress than their less hardy counterparts.[11]

Resilience

Effective combat leaders must be resilient themselves, and also must be equipped with the skills and motivation to build resilience among their followers. We have addressed resilience and related constructs at length in other chapters. The Army's Comprehensive Soldier Fitness program exemplifies a train-the-trainer approach to building resilience.[12]

But individual leaders can also be taught about the fundamental cognitive and emotional elements of resilience, and ways of imparting those skills to soldiers. Here is where future military psychologists can play an important role. Rather than leaving the leader to his or her own devices in developing resilience, psychologists may identify and empirically validate methods that small unit leaders can use on a daily basis to instruct subordinates in becoming more resistant to the adverse consequences of combat stress.

This is a lot to ask of military leaders. They must also be tactically proficient in all aspects of their job. But sustaining a motivated and combat ready unit is arguably the most important chore of any military leader's mission, and acquiring and applying strategies to achieve this end may be more important to achieving mission success—in the long run—than simple technical competence. Wars are fought by humans using a variety of tools (weapons, vehicles, command and control systems, etc.). For too long the military has emphasized these tools over the strength and psychological well-being of its war fighters. Psychologists can effect a significant increase in combat power and mission effectiveness by giving the military the methods it needs to best empower its people.

Influencing Others When People are in Harm's Way

Much of this discussion has focused on emotional resiliency, but military leadership scholar Stephen Zaccaro suggests that in order to build resilience military teams must be developed to enhance cognitive and social resiliency as well.[13] The military has long recognized the importance of both of these in preparing soldiers for war and has developed a variety of ways to achieve this end. However, this has largely been accomplished without the benefit of systematic psychological research. Comprehensive Soldier Fitness represents an important first step in addressing this, but as with emotional resilience it does not focus on training leaders with these skills. Zaccaro suggests that to influence others when people are in harm's way, the military must devise ways of systematically building resilient teams. It is just as important to focus on the collective group or team as it is to focus on individual resilience.

Zaccaro discusses several aspects of team resiliency—or, as he calls it, viability. These are cohesion, trust, and collective efficacy. Military trainers must devise ways to build and maintain these team characteristics, and the breakdown of any one of the three can prove catastrophic. Further, team resilience can be broken down into three subcomponents. These are cognitive resilience, social resilience, and emotional resilience. Cognitive resilience can be built through training and overtraining, realistic mission rehearsals, practicing effective communication strategies, and frequent and accurate feedback.

Social resilience can also be built. In demanding training soldiers learn to rely on each other, recognize who is most skilled in certain tasks, and learn who to turn to for support. This engenders collective efficacy, trust, and competence. Getting to know each other's families, engaging in recreational activities as a unit, and in general sharing a lifestyle, as Tom Kolditz suggests, is critical. In the old days, soldiers would go to "beer call" together and unwind. The modern military's near prohibition on alcohol use no doubt has saved lives, but it has also resulted in the loss of a time-honored way that soldiers bond with each other.

Emotional resilience can be built in many ways as we saw earlier in our discussion of Comprehensive Soldier Fitness. Whatever strategies are used, Zaccaro emphasizes they should include predeployment, deployment, and postdeployment efforts. In an era of persistent war, efforts to maintain emotional resilience must be ongoing.

Morale is also critical to influencing others. It is bad enough to have poor morale in a normal civilian job, but imagine how damaging poor morale is to military organizations. Colonel Brian Reed, now an army combat brigade commander, suggests several ways that in extremis leaders may increase morale in their units.[14] Leaders must (1) take charge, (2) project a sense of control, (3) give direction, (4) inspire subordinates, (5) empower subordinates, (6) maintain unit integrity, (7) share danger and hardship, (8) ensure the unit continues to grow and improve, and (9) communicate and live shared values. Leaders must also be above reproach in their personal conduct, put the welfare of their soldiers ahead of their own well-being, and always be honest and open with their subordinates.

Zaccaro mentioned the importance of cognitive resilience. James Merlo, a former army deputy brigade commander in Iraq, maintains there are many challenges to making effective decisions in combat.[15] Mortal threats and combat stress affect the perception of space and time, and change how attention is allocated, and may make cognitive processes more rigid. In combat, there are often environmental barriers to good decision making. Nighttime, fog, and complex terrain reduce situational awareness. Sometimes organizational factors hinder decision making. Poor communication channels create blind spots in the organization, and impair the ability of the leader to know what is occurring or to adjust to changes in the operational environment. For Merlo, effective decision making is crucial to influencing others in a crisis. And the solution to establishing and maintaining this ability in leaders resides in intense and regular training in conditions that simulate combat. This can be in the field or in virtual simulators. In either case, both the leader and his soldiers must train together regularly so that they can think more freely and accurately when combat begins.

Leveraging the Organization

A coherent approach to developing leaders who can perform competently and inspire others in military operations may benefit from a conceptual model that informs how to build and structure an organization. Pat Sweeney and I describe such a model, depicted in Figure 8.1.[16]

This model describes the interplay between individual factors such as self-regulation, motivation, and self-awareness within the overarching context of the social and organizational climate of military units. The central concept is that organizational and personal factors contribute to the formation of a leader's worldview. This worldview is then instrumental in influencing how the leader perceives and makes meaning of

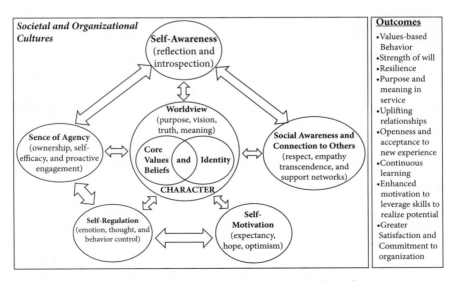

FIGURE 8.1 A Holistic Development Model for Dangerous Contexts Leaders and Organizations. (Image courtesy of Patrick J. Sweeney and the Naval Institute Press.)

events, how he frames the challenges of combat to his subordinates, and thus embodies the formation of both the leader's and followers' core values, identity, and character, as represented in the center of Figure 8.1. We believe that character is the foundation for effective in extremis leadership. Leaders who define themselves by their core values—if these values are congruent with those of the military—are more effective:

> Such people use their values to define themselves; that is, they become their values. One reasonable assumption is that the core values of an individual who chooses to work in dangerous contexts would tend to have values in line with an organization operating in such an environment and espouses values of duty, service, integrity, loyalty, courage, and respect. Being a person of character is an integral part of the identity of dangerous context leaders and "warriors"...who take on tough challenges, place duty first, never accept defeat, never quit, and never leave a fallen comrade on the battlefield.[17]

Worldview, self-awareness, sense of agency, self-regulation, self-motivation, social awareness and connections to others, and societal and organizational cultures are all things that can be developed both individually and collectively. Military psychologists will play a key role in developing a better understanding of these factors, and formulating ways of shaping and strengthening them among military leaders and followers.

IMPLICATIONS FOR NONMILITARY SETTINGS

It is worth noting that the military is not the only organization that puts its members into harm's way. Police officers and firefighters routinely find themselves working and leading others in dangerous situations. Moreover, there are many different types of dangerous situations. Dealing with an act of nature, such as an earthquake or hurricane, may require different skills and result in different actions than responding to threats caused by intentional acts of other people, such as a mass shooting.

Psychologist Donald Campbell provides a useful taxonomy of dangerous environments.[18] As you can see in Table 8.1, dangerous environments may vary on four dimensions.

First, with respect to *genesis*, dangerous environments may be adversarial or nonadversarial in nature. Nonadversarial situations would include natural disasters such as floods, fires, and earthquakes. Adversarial situations are those with an active threat, such as an active shooter situation on a college campus. Second, dangerous contexts may vary with respect to *aggressiveness*, ranging from passive (e.g., a worksite incident) to dynamic (e.g., executing a search warrant).

Table 8.1 Psychological Taxonomy of Dangerous Environments (Table courtesy of Donald J Campbell, adapted from "Leadership in Dangerous Contexts.".)

	Genesis Nonadversarial				Genesis Adversarial			
	Aggressiveness: Passive		Aggressiveness: Dynamic		Aggressiveness: Passive		Aggressiveness: Dynamic	
	Intensity: Lower	Intensity: Higher	Intensity: Lower	Intensity: Higher	Intensity: Lower	Intensity: Higher	Intensity: Lower	Intensity: Higher
Expectation: Sporadic	Work Sites Accidents: (construction)	Work Sites Accidents: (nuclear plant)	Disaster Areas Contained: (flood)	Disaster Areas Uncontained: (wildfire)	Police Work Traps: (bomb disposal)	Military Work Traps:(IED)	Police Work Traditional: (traffic stop)	Police Work Undercover: (drug deal)
Expectation: Chronic	Locations Extreme: (desert)	Locations Extreme: (space station)	Environments Training: (escape/evade)	Environments Training: (SEAL/Ranger)	Buffer Zones: (obstructions)	Buffer Zones: (minefield)	Military Work Limited (border dispute)	Military Work General (battlefield war)

Intensity represents another dimension of dangerous environments. Lower intensity situations are those with perhaps a remote potential of escalation to immediate threat to life. A police officer conducting a routine traffic stop is an example. While at some level the officer must be prepared for the possibility of a deadly threat, the vast majority of such traffic stops occur without significant incident. Highly intense situations are those where danger is present, powerful, and imminent, such as rescuing a child from a burning vehicle.

Finally, *expectations* on the part of the in extremis leader vary along the dimension of sporadic to chronic. Dealing with a sporadic threat—such as a workplace accident—invokes different psychological responses and requires different psychological resources than dealing with chronic threats. As a police officer, I often responded to intense incidents involving high degrees of physical danger, but they usually lasted only a very short time and might not occur again anytime soon. Given the nature of twenty-first-century war, a deployed soldier may face an elevated threat of mortal danger for months at a time. As a police officer, my nervous system had time to recover from such incidents, but a soldier's nervous system must remain prepared to react to danger with little respite for long periods of time.

Although genesis, aggressiveness, intensity, and expectation are depicted in Table 8.1 as being dichotomous, in fact they lie on a continuum from one extreme to another. Thus, as psychologists strive to understand leadership in dangerous contexts, they must include in their research a wide array of variations among these dimensions in order to tease out interactions. It may be, for example, that leadership situations that are nonadversarial, passive, low intensity, and sporadic may be more akin to the sorts of leader's requirements found in industrial or commercial settings. And leadership tactics that are well suited for adversarial, aggressive, intense, and chronic settings may be uniquely different from situations where one or more of the dimensions are less extreme. It is up to military psychologists to provide answers to the question of what represents the optimal leadership approach to these different types of situations.

SUMMARY

The iconic World War II general George Patton was quoted as saying, "Wars may be fought with weapons, but they are won by men. It is the spirit of men who follow and of the man who leads that gains the victory."[19] I suspect General Patton didn't give much thought to psychology as a science, much less to it as a difference maker in war. But his quote underscores the fundamental importance of understanding human nature in the leading of soldiers in combat. I suspect that almost any contemporary military leader would agree that without a properly led and motivated force, weapons—however powerful and sophisticated—are insufficient to win the day.

Psychology has the analytic and computational tools to radically advance the science of in extremis leadership. Over the next decades, military psychologists will develop a much stronger understanding of leadership, based on systematic analysis, not anecdotes and conjecture. This science-based approach to leadership will, in turn, result in a better-led, much more combat effective force. Combined with the most sophisticated and powerful technologies in the history of war, such a military may be virtually undefeatable. Let us hope that the politicians who commit the military to war do so in a restrained and thoughtful manner, and not employ our forces for economic or political gain. Only in wars perceived as just will soldiers give their all to achieve victory.

NOTES

1. Curtis Brown, ed., *Roots of Strategy*.
2. Thomas A. Kolditz, "Leading as if Your Life Depended on It," 160–87.
3. Thomas A. Kolditz, *In Extremis Leadership*.
4. For a more in-depth discussion of positive psychology and its relation to the military, see Michael D. Matthews, "Toward a Positive Military Psychology," 289–98.
5. Paul T. Bartone, "The Need for Positive Meaning in Military Operations," 315–24.
6. Patrick J. Sweeney, Vaida Thompson, and Hart Blanton, "Trust and Influence in Combat," 235–64.
7. Patrick J. Sweeney, Michael D. Matthews, and Paul B. Lester, eds., *Leadership in Dangerous Situations*. In designing this book, we paired scholars with practitioners on most chapters. Our intent was for the scholar to provide a conceptual and theoretical basis for leading in dangerous situations, and for the practitioners—combat, law enforcement, and fire department leaders, for example—to provide insight into how these concepts and theories play out in real organizations.
8. Paul B. Lester and Cynthia Pury, "What Leaders Should Know about Courage," 21–39.
9. Probably the most robust contemporary scientific model of stress involves the concept of allostasis. See Bruce S. McEwen, "Allostasis and Allostatic Load Implications for Neuropsychopharmacology," 108–24.
10. James Ness et al., "Understanding and Managing Stress," 40–59.
11. See the discussion of hardiness in Chapter 2 of this book.
12. For a clear discussion of how CSF fits into the larger picture of positive psychology, read Martin E. P. Seligman, "Army Strong," 126–51.
13. Stephen J. Zaccaro et. al, "Building Resilient Teams," 182–201.
14. Brian Reed et al., "Morale: The Essential Intangible," 202–17.
15. Joseph W. Pfeifer and James Merlo, "The Decisive Moment," 230–48.
16. The basis of this model is described in Patrick J. Sweeney and Michael D. Matthews, "A Holistic Approach to Leading in Dangerous Situations," 373–91.
17. Sweeney and Matthews, "A Holistic Approach to Leading in Dangerous Situations," 376.
18. Donald J. Campbell, "Leadership in Dangerous Contexts," 167.
19. Edwin M. Sumner, *Sept–Oct 1943 The Cavalry Journal*.

9

LEADING OTHERS IN THE DIGITAL AGE

Principles don't change—but battlefield execution in accordance with these principles has changed drastically. Soldiers don't change—but the tools of their trade, the modern weapons systems that are flooding into the inventory, are changing in a revolutionary way...As we look at the mistakes in generalship over the past 100 years, the common theme is that the general did not understand the technology of his time or, as they say, he elected to fight the last war rather than the one he happened to be in. A General must be versatile enough to take into battle the existing technology of whatever moment in time he is called upon to fight. The job of a general is to be a battlefield leader, a tactician, a logistician, a commander who readies his force for battle with enlightened training and leads it into the fight with inspirational tactical judgment and a deep understanding of soldiers.
General John Rogers Galvin[1]

General Galvin's comments in 1985 exemplify a key ability of strategic military leaders—to anticipate and to respond to change. General Galvin graduated from West Point in 1954, and went on to command the army at every echelon, including Supreme Allied Commander, Europe. After retiring from the army, he became the Dean at the Fletcher School of Law and Diplomacy at Tufts University. He is a true soldier-scholar.

General Galvin saw a lot of technological change during his time on active duty. He entered West Point just five years after the end of World War II and did not retire until after the cessation of Gulf War I. In between our nation fought the Vietnam War and various other smaller engagements around the globe. The basic nature of soldiers may not have changed during his tenure, but military technology experienced a revolution during this period. Airplanes and ground vehicles became faster and more complicated, and could carry a greater variety of weapon systems. Analog displays were replaced with digital ones. Intercontinental ballistic missiles could bring total destruction anywhere in the world within 30 minutes of launch. Slow and hierarchical command and control gave way to almost instant digitally based communication.

If you are of a certain age you can relate to this change. A generation ago you may have had a rotary telephone. Computers were mysterious devices hidden in the deep recesses of laboratories, and you had to have special training to use one. You probably got your news from watching Walter Cronkite or other evening news show. You may have received both a morning and an afternoon paper. Most educated people subscribed to a weekly news magazine. If you worked in a large and complex organization, you seldom had direct communication with your company's president.

Think how things have changed. Most people do not even own a traditional telephone anymore. I have one sitting on my desk, but it can go days and sometimes weeks without ringing (of course it doesn't really ring, per se; rather it makes a not quite pleasant digital sound). Now, because of cell phones, I am within reach of my phone 24/7/365. If there is a satellite within range, I can no longer dodge calls. My wife recently bought our first smart phone. It is well named, except we seldom use it as a phone. Now we can text, post messages on social media sites, and otherwise avoid direct human contact. I routinely use three or more computers in my job at work, and have two more that I use every day at home. Through e-mail, not only do I get regular messages from my immediate chain of command, I get them from the highest echelons of command. We even get e-mails from the President of the United States, especially during election years.

What does this mean for you, in your work? Many of us routinely spend a good chunk of each day reading e-mails and deciding which ones require our immediate attention, and which ones we can ignore. Your boss can reach you instantly, day or night, on weekends and even when you are on vacation. Many of us work in jobs that allow full access to the tools and information needed to work—through the ability to log in to your work computer from outside the workplace—so there is no excuse not to respond to questions and demands when you receive them. You may have become so used to this state of affairs that you feel anxiety if for some reason you are "off the net." What causes more angst in an office than the e-mail or entire computer system going down?

These changes have tremendous implications for how military leaders—its general officers and senior civilians—communicate, manage, and lead both in times of war and in times of peace. I am focusing my discussion on the impact of changes in command and control technology on the very nature of military leadership of the twenty-first century. These effects apply to formal communication and leadership in military units, but also affect how individual soldiers communicate with each other as well as friends and family. These changes will drive radical alterations in the traditional military chain of command, and continue to have profound psychological and social effects on our military members.

NEW COMMUNICATION AND CONTROL TECHNOLOGIES
CHANGE THE NATURE AND STRUCTURE OF LEADERSHIP

Until relatively recent times, military communication was slow and uncertain. In fact, the military chain of command as we currently know it evolved to facilitate the transmission of orders and other key information down and across organizations through voice communication. An army division commander, usually a two-star general, could issue a direct verbal order to his three or four brigade commanders (each a full colonel). They in turn could call their three or four battalion commanders (lieutenant colonels) together, and pass that order to them. Each battalion commander was responsible for (you guessed it) three or four company commanders (captains), who then relayed the information to three or four platoon leaders (lieutenants) who could then distribute the information directly to their 30 or so soldiers, or let their squad leaders (responsible for 10 or so soldiers) do so.

It is obvious that this hierarchical structure of command and communication, while necessary prior to the advent of modern communication technologies, could be painfully slow and prone to errors. It is not unlike when you are waiting behind a line of cars at a red light. Perhaps you can see the light turn green, but it may take quite a while before you can actually start moving, and you might not get through the intersection before the light turns red once again. An order is issued and at every echelon along the chain of command there is some degree of delay. Because large armies are widely dispersed across a battlefield, further delays are inherently unavoidable as information winds its way down and across the organization.

Moreover, such communication is error prone. Almost everyone is aware of the childhood game wherein one child whispers a simple sentence to the child seated next to him or her, and by the time this message is relayed across several children the original meaning may be totally lost. Psychologists, of course, have studied the relationship between organizational structure and communication, and observe similar phenomena. Despite various methods to preserve the meaning and integrity of a given message, distortions are common. Combined with delays, this can prove problematic in military command and control.

In Chapter 4 we talked about the OODA loop. Although this model was developed to describe fast-paced decisions required by fighter pilots, the concept can be generalized to chain of command communication. Just like a fighter pilot, a strategic military leader must also *orient, observe, decide,* and *act.* An organizational structure that allows the commander to do this faster and with more accuracy has a decided military advantage over a slower and less accurate opponent. Even if both the friendly and opposing forces are equal in the quality of decisions made, the organization that can execute the fastest will usually win.

Here we can turn again to General Galvin's comments regarding the pitfall of being ready to fight the previous war, not the present one. The US military first learned this lesson the hard way in Vietnam, where traditional command and control hierarchies—used in the global wars of the first half of the twentieth century—were applied in a different military context. Despite our inordinate advantage in raw military firepower, the North Vietnamese and Viet Cong were often several steps ahead of American decision making. By the time we identified a viable target, developed a plan of attack, and executed the attack, the enemy was often long gone. They had gotten inside of our OODA loop.

The Role of the General Officer in Traditional Command and Control

In a highly hierarchical chain of command, the role of the general officer and other senior commanders is to focus their cognitive and attentional resources on strategic matters. As one descends the chain of command, at each lower echelon the focus becomes narrower and more specific. There are several layers of command between the soldiers who do the fighting and the general who directs the overarching operation. For example, an army division commander—a major (two-star) general—has at least five layers of command between him and the soldiers who make up the individual squads that conduct day-to-day operations.

In this command structure, the division commander should not spend any mental resources on what is occurring at the squad level. In fact, a "two-up, two-down" command focus characterizes a commander's attention at any given echelon. According to this model, a battalion commander would try to share a mental model of the operation with both the brigade and division commander while at the same time maintain an active role in the tactical operations of his subordinate companies and, under some circumstances, their constituent platoons. In day-to-day operations, the battalion commander would primarily be occupied in executing operational orders issued through his brigade commander, which would in turn be executed by his own company commanders.

From a cognitive psychology standpoint, there are several implications concerning the requisite cognitive skills needed to be effective as a function of echelon of command. Broadly speaking, military tasks can be labeled as tactical, operational, and strategic in nature. Tactical refers to small unit missions such as patrolling an area, maintaining security over a valued target, or accomplishing the myriad day-to-day tasks necessary to maintain the unit. A platoon leader will devote the majority of his cognitive abilities to completing these sorts of tasks. The operational level of focus is broader and encompasses larger goals and objectives. Operational tasks require greater coordinated communication with other units, above, below, and laterally in the organization. An operational task may be to train local police so they can conduct law enforcement duties and quell insurgent activity within a given state or province. This operation can be decomposed into subordinate tactical missions, but the commander must have a larger picture of the overall mission than the commander of a small tactical unit. As a general rule, operational tasks are executed by battalions and perhaps brigades, although this may vary with respect to the specific mission. The strategic level refers to the achievement of overarching goals, such as winning the war, preparing a nation-state to self-govern, and so forth. These are precisely the tasks that general officers are trained and developed to undertake.

The decision-making skills and strategies that define an effective strategic leader are quite different from those that enable a platoon leader to make accurate decisions. The strategic leader has more time, a larger staff, is usually not under the pressure of receiving direct fire, and may be working in a relatively comfortable physical environment, free at least from the stresses of cold, heat, and precipitation. This leader must have more elaborate mental models of the overall mission in order to decide among alternative courses of action. He or she must learn to direct attentional resources to a plethora of information sources and show flexible and adaptive thinking. They must have what cognitive psychologists refer to as high metacognitive skills; that is, they must be able to coordinate and self-regulate the components of decision making—perception, attention, working memory, and long-term memory—to render effective decisions.

While small unit leaders certainly invoke similar cognitive processes, their decision making may differ in substantive ways compared to the higher-echelon commander. Because they must often make snap decisions, they may rely more on pattern-matching of scripts to particular situations. They may not always be able to verbalize why they made a certain decision. They may need to focus attention more on the immediate environment rather than the abstractions of a strategic goal or policy. Their mental models must focus on the exigencies of tactical missions. Their information input may be restricted to what they can personally observe or obtain from verbal reports from members of their own unit.

The current hierarchical military command and control structure has worked relatively well. But revolutionary changes in technology may necessitate a change in this structure. This will result in substantive changes in the role of military leaders, and psychology will play an important role in influencing how these changes evolve.

Military Command and Control Today

Stop and reflect on how the command structure described above has changed. The most vivid example is the televised image of the US Commander in Chief, surrounded by the nation's military and intelligence services leaders plus the Secretary of State, monitoring the Navy Seals operation that resulted in the death of Osama bin Laden. Imagine President Lincoln monitoring the Cemetery Ridge battle at Gettysburg from the White House using today's technology rather than the telegraph. Unthinkable.

The circumstances of the bin Laden raid were unique and resulted in significant strategic consequences. I am not suggesting that the commander in chief makes a habit of keeping tabs on small unit missions, or that he was incorrect in doing so. But the sheer ability to do so raises significant questions about the very nature of command and control. In the past, a strategic leader would have found it virtually impossible to track and possibly intervene in small unit operations, even if they desired to do so. But with modern digital technologies, available planet wide through satellites, and coordinated by fast and intricate systems, it is quite possible that commanders can easily step outside of the old two-up and two-down doctrine.

Figure 9.1 shows a tactical command post at a military installation in Afghanistan. While the physical layout of command posts will vary with echelon, service, and function, each contains a host of highly sophisticated technologies. Indeed, the old acronym for command post (command and control, or C2) has changed and now is C4ISR, which stands for command, control, communications, computers, intelligence, surveillance, and reconnaissance. The new acronym alone suggests a transformation

FIGURE 9.1 US Army Tactical Command Post in Afghanistan. (Photo courtesy of Russell Lemler.)

of the function of the command post in the current era. Carefully choreographed displays of digital screens present near real-time information to the commander and his staff on all aspects of the military operation. This includes the status of friendly forces and similar information about enemy forces, gathered by human intelligence, satellites, and battlefield sensors. Information on terrain and weather are available instantly. The commander has the ability to communicate directly with other units via a variety of technologies, including voice and text. The commander can call or participate in high-level staff meetings, face-to-face, with others worldwide, through secure teleconferencing technologies.

The C4ISR doctrine reflects the technologies available on the modern battlefield. These systems can be found on the ground, in what the army calls Tactical Operations Centers, or TOCs, and also onboard ships or in aircraft. The expansion beyond the traditional C2 model emphasizes the role of computer technology that provides the commander and his staff vastly improved communications through enhancing both the speed of delivery of orders and other information and the breadth that can be shared. The intelligence component—always critical to military success—is improved through the ability to quickly update and portray near real-time information about the enemy. The reconnaissance function reflects the capability of using various space-based, airborne, and land-based sensors to provide factual information about the enemy, in essence extending the commander's ability to sense and perceive the enemy directly.

You may be wondering what role psychology plays in such technologies. The answer to this question was provided to me personally a few years ago by General Eric Shinseki, who had recently retired from the position of Chief of Staff of the US Army. My department at West Point was fortunate to have General Shinseki as its inaugural Class of 1951 Leadership Chair in Behavioral Sciences. I arranged a meeting between General Shinseki and cadets enrolled in one of the engineering psychology courses I was teaching that term. General Shinseki pointed out to the class that there was a lot more psychology in C4ISR than pure technology. Commanding and controlling others has always required a strong understanding of human nature and motivation. Communication is by its very nature a social psychological phenomenon. Intelligence requires sophisticated mental models and reasoning skills—to make sense of disparate and often contradictory information. Human factors engineers play a critical role in the computer piece of C4ISR by designing displays to organize and display complicated arrays of information in ways that are congruent with how humans perceive and process information, and to avoid overloading the commander's sensory, perceptual, and memory abilities. Perhaps that leaves surveillance and reconnaissance as relatively nonpsychological in nature, but these contribute directly to the commander's mental model of the battle space, hence these too are psychologically relevant.

Because of their bulk, contemporary C4ISR systems are generally designed for battalion and higher echelons of command. They are staffed by a host of officers and enlisted specialists who assist the commander in his duties. In the future, C4ISR technologies will be miniaturized to the point they can be mounted on individual soldiers. Prototypes for such systems have already been developed and field tested. One such system, discussed in detail in Chapter 4, is the Army's Land Warrior System, which gives individual soldiers the command, control, and communication capabilities currently available in the TOC at the battalion level or higher.

Future Command and Control Technologies

Future iterations of individually mounted C4ISR systems may include even more revolutionary changes. We will look into this more fully in Chapter 11, but Figure 9.2 depicts a representation of the Marine Corps' Space Trooper concept. Besides a more sophisticated version of the Land Warrior System's weapon and supporting systems, this and similar systems under development may include a battle uniform that regulates the soldier's temperature and monitors vital health indicators such as energy expenditure, heart rate, EEG, and similar physiological measures. Using nanotechnology, the

FIGURE 9.2 The Marine Corps Space Trooper Concept. (Photo courtesy of Algol; Shutterstock.)

suit may be chameleon-like, changing color and texture gradient to camouflage with changing environments.

The psychological ramifications of contemporary and future C4ISR systems are profound, and psychologists will play a key role in their design and training soldiers to use them. Here are some examples of how these technologies intersect with psychology.

Technologies Will Change the Social Structures of Combat Operations

It is widely acknowledged in the military that soldiers fight for each other, and not so much for abstractions such as freedom and democracy. The latter justifications of war are more often voiced by politicians who try to sell the rightness of a given war to the public. Soldiers fight for their buddies, who traditionally they could literally reach out and touch, or at least see and maintain direct voice contact. Keeping soldiers together in relatively close quarters facilitated command and control, and focused more firepower on the enemy. A byproduct of this was that soldiers could motivate and comfort each other, as needed. It decreased loneliness and gave them a social outlet to express their hopes and fears.

But Future Force Warrior and similar systems give the individual soldier the firepower of a traditional squad or platoon. More importantly, the technology allows a small unit such as a platoon to effectively fight in and control the territory that may have once required a company, or even a battalion, to effectively manage.

It is this dispersal of the individual soldiers that may prove psychologically daunting. Imagine being far from home, in harm's way, and you do not even have the comfort of another soldier physically near you. This could heighten fear and anxiety and lessen a soldier's motivation to engage the enemy. The extent to which a virtual presence of other soldiers, in the form of text messages and radio communications, may ameliorate the adverse effects of dispersion are not known. Today's youngsters seem to spend more time texting each other and interacting in virtual space through gaming than they do in face-to-face play and interactions. Perhaps future soldiers, who grow up interacting in such ways, may not be disturbed by physical isolation. But my guess is that human nature itself will not change much, and that when placed in situations of mortal danger, the physical presence of others will be critical to both mission success and soldier well-being.

Psychologists, therefore, must be involved in assessing the impact of soldier dispersion and physical isolation on motivation, performance, communication, and personal adjustment. This is too important to allow Darwinian processes of trial and error to dictate how the issue is addressed. Military leaders and trainers, no matter how skilled at tactics and leadership, need the sage advice of military psychologists in developing tactics, techniques, and procedures that will support future combat operations.

Clinical psychologists can assess isolation's impact on personal adjustment. Social psychologists can conduct studies aimed at optimizing communication and social support. Human factors psychologists will help design digital systems that offset the effects of physical separation by enhancing certain aspects of the virtual space the soldiers will occupy.

The ability to communicate rapidly and horizontally through an organization may allow a change in the traditional chain of command. If the current chain of command is simply a legacy based on pre-twentieth-century command and control technologies, how can it be changed in order to improve and speed communication? There is no firm answer to this question at this point, but I submit that psychologists may be of great assistance to the military in exploring alternative command structures, reviewing their strengths and weaknesses, and developing training to help personnel operate effectively in the new structure. It may well be a disservice to the military to fight wars of the mid-twenty-first century with an organizational structure that dates back hundreds of years.

Getting the Right Information to the Right Soldier at the Right Time

I am currently reading an amazing first-hand account of the World War II battle of Peleliu. Based on notes taken during the battle by a then 20-year-old marine corps rifleman Eugene Sledge, the book gives a detailed account of classic infantry warfare.[2] Sledge relates at one point that at his level in the battle—basically small unit action embedded within a company—he and his fellow enlisted soldiers had no idea of where they were or, except in general terms, what direction they were moving. Only officers had maps.

With technologies that will equip the future dismounted soldier, the next generation of Eugene Sledges will not suffer from lack of information. Every soldier will have the same view of the battlefield as the commanding general. He will know his exact location, see icons representing enemy and other friendly locations, and have a clear picture of the commander's intent. At the same time, it will be easy for commanders at higher echelons to drill down to the level of small squad operations and observe tactical encounters in real time as they unfold. They will be able to communicate with any echelon within their command by simply typing a text message.

While these technologies are designed to improve command and control, they also raise important questions about who should receive what information, and when that information should be transmitted. First, from the standpoint of an enlisted soldier operating at the small unit level, while it may be useful to know one's exact location, it could also prove confusing and even detract from a soldier's proper focus on the immediate tactical mission. It may be necessary to place limits on where and to whom a soldier may convey information or simply "chat." Too much information flying back and forth, and up and down, could add to the confusion that is inherent in battle.

On the other side, some military leaders—like managers and leaders in all domains—are prone to micromanaging. New technologies set the occasion for the micromanagement of military operations in ways that in the past have not been possible. A high-echelon commander could violate the two-up, two-down rule and end up interfering in operations of which he may not himself have a full understanding. The psychological distance of the commander from the fight will be diminished. High-echelon combat leaders know, in the abstract, of the suffering and sacrifices made by their soldiers. But with the ability to monitor real-time activity at the squad and even individual soldier level, stress and concern for the immediate welfare of troops could affect the commander's decision making in unknown ways.

Here again psychologists can help the military design communications systems that reside in future command and control systems in such a way that they enhance this function, and not detract from it. Social and organizational psychologists need to study the impact of these emerging technologies in the military context. While there may be parallel changes resulting from technology in the civilian domain, leading and coordinating the actions of large numbers of other people under life-and-death conditions may result in second and third order effects not seen within a civilian company.

Will New Technologies Alter Military Culture and Adversely Affect Discipline?

Compared to previous eras, today's all-volunteer force enjoys the benefit of a junior enlisted force that is 100 percent literate, and nearly 100 percent educated to the level of high school (or equivalent) or beyond. Even in war zones, soldiers either own smart phones or have access to the Internet, and can stay informed of news and political events. Traditionally, soldiers were simply expected to obey their orders. They did not have access to the more generalized picture of the operation that their superiors enjoyed or easy access to general news sources.

Will tomorrow's soldiers be more likely to debate the wisdom of orders than past soldiers? Much will depend on how the military's organizational structure evolves to fully exploit these new technologies. It will also depend on training and the organizational culture in tomorrow's military. It is plausible that highly knowledgeable soldiers may expect more information and demand more in terms of explanations of their assigned mission from their commanders. This trend may be exacerbated by the increased number of highly specialized skills needed to wage future war. Cyberwarriors, no doubt well-educated, may not "snap to" as readily as a private in World War II.

This may not be a uniformly bad outcome. If one views a military organization as a single but diverse organism, consisting of a motor system, a central nervous system, and sensory systems, then figuring out ways to quickly integrate input from the individual cells and subordinate systems (soldiers and units) may improve the effectiveness and efficiency of the organization. If the commander is the organization's frontal lobe, then fully utilizing his distributed cognition resources (immediate staff, and input from soldiers and subordinate units in the field) should improve decision making. The biocognitive model of military organization, coupled with innovations in information technology, may provide a psychological model for how to improve unit effectiveness.

The ability of soldiers to communicate in real time with loved ones at home also may prove problematic. My colleagues who have deployed to Iraq and Afghanistan say that e-mail, Skyping, and similar technologies are a mixed blessing to the soldier. On the one hand, they allow soldiers to remain connected to their families and friends. While not physically present, they may still be virtually present for important events like a child or spouse's birthday. On the other hand, knowing the day-to-day challenges that one's family is facing in your absence can be distracting and lead to increased worry and anxiety. And some spouses hand off important decisions to the deployed husband or wife simply because they can. The stress of deployment and combat, while lessened somewhat by the ability to stay in close and immediate touch with family, may be magnified if the relationship is dysfunctional or the home-based spouse or family do not know how to relate and communicate with the soldier in a productive way. (You may prefer *not* to know that your dog died.)

Of course, bad social and family relations have plagued some soldiers since the dawn of time. A close friend of mine returned from combat for his mid-tour R&R (rest and recreation) only to be informed by his wife that she had filed divorce papers and was going to take their two kids and marry a mutual friend, who was divorcing his wife so this could happen. Even more vexing, the other man involved in this was not deployed. Countless soldiers have received "Dear John" letters. While the ultimate outcome may be the same, the real-time nature of family and spousal discord that is enabled through modern information technology may render affected soldiers ineffective and destroy their morale.

Whether military discipline is or is not altered by changing technology will rest in part on how psychologists help the military learn to optimally integrate and utilize new technologies. Social and organizational psychologists will lead the way in this work, and clinical psychologists may inform the military as an organization and individual soldiers on how to deal with both positive and adverse family situations. Military effectiveness hinges on morale, so it is imperative that psychologists begin to more thoroughly examine the impact of emerging information technology at both the individual and organizational level, and for both work and personal interactions.

What to Do When the Technology Fails?

I seldom think about how reliant I am on modern technologies until they fail. The server crashes and I no longer have e-mail, or there is a power outage and I lose my cable service. I enter a dead zone and lose my smart phone satellite connection. I become agitated and frustrated when I can't e-mail, check Facebook, or see my favorite television show (from the 500 channels I receive). You are probably the same way. But it is only an inconvenience. If the problem persists, you rediscover other ways of communicating or obtaining information. You may actually call someone on the telephone, or listen to the news on the radio.

But what happens when sophisticated technologies fail on the battlefield? All modern weapons systems are controlled by computers, and all current and future command and control technologies tap into a variety of digital systems, as we have seen.

Consider, for instance, the task of aiming an artillery piece on a particular target. Digital technologies allow this to happen with virtually zero degree of error. A missile or smart bomb may hit within inches of its target, despite being launched a great distance from the target. Prior to the advent of these technologies soldiers spent hundreds of hours learning how to read a map, and apply complicated mathematical formulas toward the task of placing a round on or near a target. This is not an easy skill to acquire in the first place, and it takes months of training and continued practice to keep the skill set fresh.

Our military has fought in environments where we maintain a vast technical and military superiority over the enemy. The enemy has thus far not been able to destroy or even temporarily interfere with our battlefield technologies. But a nuclear pulse, resulting from the detonation of a nuclear device high above a battlefield, would render all but the most hardened technologies inoperable. Space-based weapons could take out the satellites that are the basis for GPS information. Critical higher-echelon commander centers could be destroyed or overrun by a well equipped and well led enemy. Cyberwar could be used to hack into military systems and disable them, render them useless, or, worst of all perhaps, plant false information.

The military trains soldiers with the skills to fall back on old ways of accomplishing tasks. If digital target finding technologies fail, the tanker can in theory calculate range and trajectory using old analog methods. Commanders can fall back to radio communication or even send runners to convey information. But it is likely that once a strong reliance on the digital systems has been developed, parallel analog skills will be rusty at best and probably not functional.

Psychologists must provide the military with effective ways of training, maintaining, and retaining backup skills. Technology failure in a context where mission success and lives are on the line will also provide yet another form of stressful event, and psychologists can help provide military personnel with the emotional skills and coping strategies to deal effectively with these situations.

SUMMARY

I return to the remarks of General Galvin about the military needing to keep up with changing technology, and I extend his comments to say that military psychologists also need to keep up with changing technology. Emerging technologies have and will have significant effects on military organizational structure and command and control. Psychologists can help the military get ahead of these changes, and prepare it to change and evolve in ways to sustain and improve performance, discipline, and morale.

Leadership depends on the commander instilling an organizational climate of competence, trust, and esprit de corps. The hierarchical and authoritarian leadership model that worked well for wars of the twentieth century and before will not lend itself to either the nature of future wars, or the technologies used to fight them. The military faces a revolution of leadership theory. The military cannot afford to let Darwinian processes dictate the course of these changes. Psychologists and other behavioral and social scientists must be coopted into a systematic and proactive effort to tailor military leadership to the demands, technologies, and conditions of future war. Victory depends on it.

NOTES

1. US Department of the Army, *Leadership Statements and Quotes*, 8.
2. Eugene B. Sledge, *With the Old Breed*.

10

THE TWENTY-FIRST-CENTURY PATTON

My mental faculties remained in suspended animation while I obeyed the orders of the higher-ups. This is typical with everyone in the military.
Major General Smedley Darlington Butler[1]

Major General Butler, the most decorated officer in the history of the marine corps at the time of his death in 1940, was quite a warrior. One of only 19 men to twice receive the Medal of Honor, he led soldiers in combat in Europe, Mexico, the Caribbean, and Asia. Later in life he became an outspoken critic of how the United States used its military forces. As you can see in this quote, he was also critical of the prevailing style of leadership common to the military of the early and mid-twentieth century. The purpose of this chapter is to explore the leadership styles that will be needed in wars of the twenty-first century. As we shall see, they will not reflect the blind obedience to authority criticized so succinctly by General Butler.

In the last chapter we looked at the ways in which technology might influence military leadership. Changes in technology will drive changes in communication, persuasion, compliance, and obedience to authority. Psychologists must work hand in hand with the military to help shape its future organizational structure to reflect these changes. These technological changes are occurring in the context of social changes that may require military leaders—from the most senior general to the most junior sergeant—to learn new ways of leading others.

I suspect that those who have not served in the military may have somewhat unrealistic views of day-to-day leadership in the military. Movies often depict officers barking orders to subordinates, who in turn salute smartly and rush to carry out those orders. Military leaders are often depicted as being brash and curt in their interactions with their soldiers. They appear to have a great social distance from their enlisted soldiers. They may be perceived as placing their careers above the welfare of their soldiers, or at least disconnected from the day-to-day lives of their soldiers.

There is some truth to these stereotypes. The hierarchical nature of the military organization combined with the rigid and time-honored rank system is designed to maximize rapid and accurate transmission of formal orders and information quickly and with minimal distortion. In combat, it is imperative that commands be communicated and carried out in the manner envisioned by the commander.

But this view of military leadership is at best incomplete. An officer or NCO who relies only on his or her legal authority to order subordinates to complete specified tasks is doomed to failure. In my years as an air force officer, I supervised several enlisted airmen, and was supervised by several different officers. I do not recall even one instance where I ordered someone to do something, nor a single instance where I was issued a direct order. Moreover, in 15 additional years working as an army psychologist with daily interactions with officers and soldiers, I have not seen or heard an order—as depicted in the movies or on television—being issued. I have no doubt that this occurs. The issuance of direct orders may be more frequent and necessary in a combat setting; one could hardly confuse my military service as combat, to be sure, but my officer colleagues at West Point, all of whom have served in combat either in Afghanistan, Iraq, or both, also say that Hollywood type orders are seldom necessary or appropriate. Effective leadership requires much more than positional authority and a loud voice.

CONTEMPORARY MILITARY LEADERSHIP DOCTRINE

Contemporary military leaders spend their careers developing and honing the skills they need to effectively manage the behavior of others in trying and often dangerous contexts. At West Point, while cadets complete an academic major similar to those available at other institutions of higher education, leader development is interwoven into all of their activities. For 47 months, in the classroom, on the athletic fields, or while engaged in learning specific military tasks, cadets are taught fundamental principles of leadership. Moreover, through a cadet chain of command that parallels the organization of a regular army brigade, they are given the chance first to learn to follow (as plebes) and ultimately command others (as seniors, through various command and staff positions). They are *not* taught to bark orders, drop subordinates for pushups, and to otherwise harass, intimidate, or haze others.

The curriculum at other service academies, ROTC, and officer training and candidate schools is similar. In 12 weeks of training at Air Force Officer Training School, I learned that the best officers leverage their competence, trust, and knowledge of human behavior to get others to complete tasks. We learned the do's and don'ts of leadership—not simply the chain of command and the proper tone of voice or posture in issuing verbal orders. Like other commissioning sources, there were substantial blocks of education on leadership, human behavior, and motivation.

Each service—the US Army, Air Force, Navy, and Marine Corps—has its own culture of leadership and specific leadership doctrine, but the similarities are more noticeable than their differences. It is worth reviewing leadership doctrine from the lens of a psychologist to illustrate what contemporary military leadership entails. I will use US Army Field Manual 6-22 (*Army Leadership: Competent, Confident, and Agile*) as the basis for discussion, but the interested reader can easily find the corresponding doctrine of the other services through a simple Internet search.[2]

FM 6-22: The Basics of Leadership

The core philosophy of army leadership doctrine is captured in the words "be, know, do." These simple words are based on hundreds of years of practical leadership experience as well as on principles of modern leadership theory. They provide a reference point for leader development and leader practices.

"Be," Internalizing the Values

Effective army leaders must internalize values and attributes that are congruent with the organization's mission and purpose. What, exactly, are these qualities? The army specifies seven fundamental values that are critical to leadership. These values are shown and defined in Figure 10.1. The navy, marine corps, and air force each have their own list of similar values. Important for all soldiers, it is especially critical that leaders exemplify these values. A leader must be loyal to his or her subordinates in

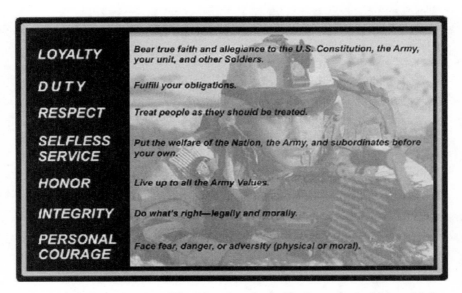

FIGURE 10.1 The Seven Core Army Values, from *Army Leadership: Competent, Confident, and Agile*, Field Manual 6-22. (Image courtesy of Headquarters, Department of the Army.)

order to receive loyalty from them in return. Social learning theory tells us that modeling is a powerful way to influence others. The army leader—officer, warrant officer, or NCO—who is loyal, is devoted to duty, respects those above, below, and next to him, subjugates his own welfare to that of the greater good, consistently adheres to his values, is honest both legally and morally, and has the fortitude to face danger and adversity in its various forms, is poised to inspire similar behavior in followers. Especially in combat, a leader or soldier who is lacking in even one of these values may pose not only a risk to those around him, but also to the mission itself.

Contemporary theories of human character suggest there are as many as 24 character strengths—internalized and stable sets of values and beliefs—that are valued in all cultures.[3] Each of the army seven values is represented among these strengths. Evidence shows that officer candidates at West Point and other military academies show character strength profiles that reflect the army values.[4] Moreover, when asked what character strengths are most important in dealing with the adversity of combat deployments, army officers rate—from the 24 character strengths common to all people—those that pertain to the army seven values as being most important in battle.[5]

"Know," The Importance of Competence

It is not enough to be a leader of high character. A leader must also be an expert in the tactics, techniques, procedures, and doctrine of his craft. The leader must also know how to effectively manage financial and technical resources. A leader must possess high social intelligence, be able to observe and recognize the tendencies and needs of people, and know how and when to influence them.

Patrick Sweeney's study of army leadership during combat operations in Iraq in 2003 underscores the importance of competence.[6] He found that subordinate soldiers looked first and foremost at the professional and technical competence of their leaders. An incompetent leader could not be trusted. But technical competence was not enough. In terms of logic, it is a necessary but not sufficient condition to inspire trust among followers. All three components of the Army Be-Know-Do model had to be present.

"Do," Actions Matter

This third component of effective leadership simply refers to the necessity for leaders to apply what they know and manifest in their daily actions who they are. That is, they must display the army values and demonstrate competence in their daily actions. The leader who says he is a selfless servant but manipulates his organization and its members for personal gain cannot be effective. Soldiers are quick to identify an officer who

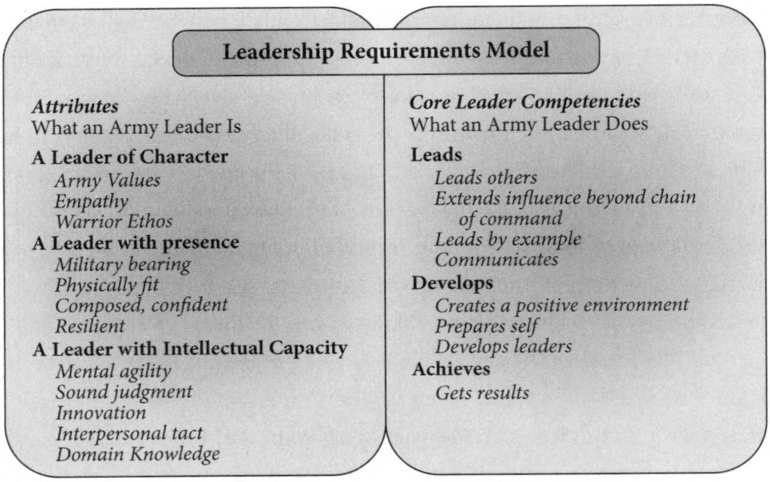

FIGURE 10.2 The Army Leadership Requirements Model, from *Army Leadership: Competent, Confident, and Agile,* Field Manual 6-22. (Image courtesy of Headquarters, Department of the Army.)

puts his or her personal promotions and achievements above the good of the organization and the mission. A lieutenant newly assigned to a combat platoon, who is overly eager to lead his platoon into combat, may be perceived as valuing personal glory over the good of the unit.

Figure 10.2 summarizes the core attributes and competencies that army leaders are expected to possess. Officers and NCOs must internalize and display character, have a command presence (be physically fit, composed, etc.), and be intellectually prepared for leadership. They also must lead (by example, effective communication, etc.), develop others, and be able to achieve results.

QUALITIES OF FUTURE ARMY LEADERS

Leadership doctrine, as reflected in FM 6-22, provides a general framework for military leadership. It clearly shows that effective leadership requires far more than placing someone into a position of command and then simply empowering them to pursue a task or mission by ordering people around. Military leadership depends on the ability to influence, inspire, and guide others rather than on what social psychologists call obedience or compliance.

A MODEL FOR FUTURE MILITARY LEADERSHIP

Contemporary leadership doctrine is vague on the specific individual and collective traits, skills, and attitudes that will be needed for leaders in future conflicts. Nor is the scientific basis of current leadership doctrine always clear. In the following pages I discuss the traits, abilities, and skills that will become ever more important for military leadership in the future. Some aspects of military leadership are timeless, and I am not suggesting a wholesale change in how leaders are selected and developed. But in the changing cultural, political, and technological nature of future war, these factors may rise in importance.

Trust, the Cornerstone for Military Leadership

The notion that trust is crucial for military leaders is not new, but the emerging psychological science of trust informs us how to set the conditions for greater trust of military leaders and, in turn, their ability to trust peers and subordinates. Patrick Sweeney describes the IROC model of trust development, shown in Figure 10.3.[7] This model emphasizes the key roles of (1) individual credibility, (2) interpersonal relationships, (3) organizational climate, and (4) context.

Individual Credibility

The basis for developing trust is individual credibility. This hinges on three factors: competence, character, and caring.

For future military leaders, *competence* must go far beyond the technical skill of weapons and command and control systems; it must also include political, social, and cultural competencies. The best tactical company grade officer—skilled as he or she may be at weapons systems—must also be proficient in a wide array of tasks. Traditionally, the military favored promoting officers with demonstrated tactical

Individual Credibility: The Foundation
- Competence
- Character
- Caring

Relationships Matter
- Respect and concern
- Open communication
- Cooperative interdependence
- Trust and empower others

Organization Sets the Climate
- Shared values, beliefs, norms, and goals (culture)
- Structure, practices, policies, and procedures

Context Influences All
- Dependencies and needs
- Organization systems

FIGURE 10.3 The IROC Model of Trust Development. (Image courtesy of Patrick J. Sweeney and the Naval Institute Press.)

proficiency (i.e., war fighting skills) to positions of high leadership. The army infantry officer and the air force fighter pilot had an advantage in promotion to senior ranks over those lacking in these skills. There was (and still is) no attempt to emphasize the importance of nontactical skills, much less assess, train, and develop them. The officer development system also favored officers strong in engineering, math, and science educations.

Given the importance of cultural IQ for future military leaders, the military should begin identifying and developing officers for senior leadership positions who have demonstrated competence in such skills. Psychology, sociology, and anthropology may provide a better educational bedrock on which to build senior officers than engineering or math. The military should add substantive blocks of instruction to officer and NCO development schools at all levels, from basic courses for new lieutenants and sergeants, to the curriculum in the war colleges. Psychologists will play a critical role in providing instruction and education in these settings. Military leaders who do not have or do not acquire these skills may end up doing more harm than good when deployed into war.

As we have seen, army doctrine (and the doctrines of the other services) specifies certain values that are deemed to be critical to military service. These are based on years of practical experience and reflect important traits that leaders must manifest. However, the emerging science of *character*—in the context of the changing nature of warfare—provides a basis for a more empirical analysis of positive character traits military leaders should exemplify.

Military psychologists have begun to empirically identify which character strengths, described in Chapter 4, are most important for military leaders. I have conducted research among West Point cadets now for nearly a decade, and have found that the strengths that are vital for successfully completing West Point include bravery, vitality, fairness, honesty, persistence, optimism, self-control, and teamwork.[8] Cadets high in these strengths are more likely to remain at West Point and perform successfully in academic, military, and physical fitness tasks.

But doing well at West Point is not the same as being an effective army leader, particularly at the strategic level. I have identified five character strengths that are commonly used by company grade officers to deal effectively with the challenges and adversity of combat; these are teamwork, bravery, capacity to love, persistence, and honesty.[9] This is not to say that these were the most common character traits possessed by these combat officers. Rather, these are the strengths they indicated they actually use in combat.

What is lacking, however, is a systematic and empirical strengths assessment of military leaders as a function of echelon of leadership, from small units to four-star command, with a linkage to success in those positions. Such an analysis may reveal particular character strengths or—more likely—constellations of character strengths that differentiate the successful military leaders from the average or unsuccessful leader. Once identified, methods for selecting leaders with these traits and/or developing these traits in emerging leaders could be designed by psychologists. Based on the likely nature of future war, I would expect that social intelligence and other people-oriented traits would rise in importance versus the more authoritarian traits that were instrumental to military leaders of past wars.

Also an interpersonal skill, for military leaders *caring* means more than the ability to empathize. Caring also means balancing the needs of the mission with those of one's subordinates. Followers must know that the leader will act in ways that support their best interests. It means the leader is genuinely concerned about the personal and professional development of members of his unit, from the lowest private to the second in command. Moreover, the commander will put their well-being and development ahead of his own needs.

The psychology of caring in this context requires more research, with the goal of identifying specific behavioral, belief, and attitudinal components. Military psychologists can leverage this knowledge of the scientific basis of caring into leader development and training programs at all echelons.

It is particularly essential that strategic military leaders—indeed, the Pattons of the twenty-first century—possess all three of the individual credibility components of the IROC model. Not only are these skills vital to command, they are also vital in interacting with native populations, nongovernment agencies, and other extra-organizational entities that the future four-star general must work with effectively to achieve the strategic and operational outcomes that make up his mission.

Interpersonal Relationships

This component of the IROC model has four subcomponents: (1) respect and concern, (2) open communication, (3) cooperative interdependence, and (4) trusting and empowering others. While there is some overlap with individual credibility, here the focus is on the quality of interpersonal interactions.

Respect and concern for others is critical to contemporary military leadership. Your stereotype of military leaders may not include these qualities, and perhaps they were less important in traditional, kinetic focused wars of the twentieth century. Tomorrow's military, even more so than today, will be a socially and culturally diverse organization. Leaders who embrace differences while still pursuing common goals, and who display genuine concern for the development of others, will be better able to lead a diverse military in the context of equally diverse missions.

Soldiers of the twenty-first century are not likely to be satisfied with an answer of "because I told you so" when asking about the reasons for particular orders or missions. Military leaders who are not afraid to express themselves freely and openly will garner trust and cooperation faster than those who do not reveal information. This *open communication* is directly linked to the concept of *cooperative interdependence*. No single soldier or unit possesses all of the skills and knowledge needed to successfully complete the full range of missions they are likely to be given. Individuals and units that communicate in a timely and accurate way, and share the right information to the right people at the right time, will facilitate mission success. Cooperative interdependence means that one's personal and unit success hinges on the success of other people and units. Future leaders may capitalize on advanced information technology to make this happen, but must also set the occasion for this type of command climate in order to ensure success.

Trusting and empowering others represents a norm of reciprocity. If a military leader is to be trusted, then he or she must in turn trust both their subordinates and their own chain of command. Micromanaging and authoritarian leadership styles, perhaps effective in the short run in accomplishing traditional military missions, will fail in the more complicated battle space of future war.

Organizational Climate

Sweeney maintains that organizational climate establishes the motivation and morale to accomplish complex missions. This includes *shared values, beliefs, norms, and goals.* The future military commander must bring together a diverse force and bring them together around core values. This is not new concept. Anyone who has served in the military has taken pride in being a soldier, airman, sailor, or marine. Within services, each branch has its own customs and traditions. Even individual units set themselves apart from others by having unit t-shirts, coffee mugs, and other cultural artifacts. Done well, this melds diverse individuals into a team. You may have heard there is no such thing as a former marine—once you have been accepted into the Corps, you will always be a marine.

Clearly defining organizational goals is critical. Getting everyone "to pull the same rope" builds esprit de corps. Every psychology student has read the classic research of Muzafer Sherif, who united groups of highly competitive boys simply by giving them a superordinate task to complete.

Building an overarching and uniting organizational identity and climate requires that leaders build an *organizational structure, practices, policies, and procedures* that are transparent and unambiguous. We have talked already about how organizational structure must change to accommodate advances in information technology. A flatter and faster (in terms of communication speed) organizational structure will require supporting practices, policies, and procedures that are likely to differ from those in place in today's military. Psychology will play an instrumental role in refining and implementing these changes.

Context

The military uses the acronym METT-TC (mission, enemy, terrain and weather, troops and support available, time available, and civilians) to capture the idea that every operation has its own unique characteristics. This has always been true, but in wars of the mid-twenty-first-century operational conditions may change at a stunning pace. An aid mission to assist citizens of a country devastated by a flood or earthquake may become a low-intensity combat mission almost overnight. This happened in 2012 when extremists in various Muslim nations reacted violently to a degrading portrayal of Mohammad by an American filmmaker. Embassies were attacked, the safety of American nationals in those areas was placed at risk, and military outposts throughout the region were put on high alert.

Military commanders must be hypervigilant regarding local *dependencies and needs* and adapt their *organizational systems* to optimally operate in that given context. More than anything else, future military leaders must be adaptive, and able to change course with little advance warning. Those who ignore the current METT-TC will lose trust and therefore lose combat effectiveness. Everything depends on this flexibility and adaptability. The commander must be able to see new ways of doing things, and not be afraid to quickly adjust courses when needed.

The psychology of this is quite complex. Cognitive flexibility and adaptability is studied by cognitive psychologists, but most of the research is done in sterile university-based research labs, with undergraduate students serving as research subjects. The psychologists are conducting clean experiments with good internal validity (meaning that their results can be reproduced by others), but suffer greatly in external validity. How a military leader exercises mental flexibility and adaptability in the context of an organization that has traditionally discouraged thinking outside the box, and in a setting where making the wrong decision can have global consequences, is quite disparate from a college sophomore switching learning strategies to adapt to a new sort of puzzle.

The military *organizational system* of the future must be designed to encourage and promote this type of adaptability. An organizational structure designed for wars of the twentieth century between superpowers will not support the requirements of speed, agility, and information flow of twenty-first-century war.

The IROC model represents the skills that future leaders at every level of the military will need. This review of the model clearly shows the role of psychology in training, educating, and developing these skills in leaders. This will be one of the most important contributions of psychology to twenty-first-century warfare.

Specific Leader Skills for Twenty-First-Century Wars

What follows is a brief description of a variety of skills that may help future generals and other military leaders successfully complete the wide variety of missions they are likely to face. I have identified these skills from years of experience while consulting with strategic army leaders on future doctrine and practices as they impact leadership, through working with West Point military faculty who will be among the army's future strategic leaders, and conducting extensive empirical research on factors that facilitate soldier and leader performance in challenging contexts. This list is based on my own interpretations of what future military leaders need to "be, know, and do"—this is not something to be found in any official doctrine of policy. As technology continues to evolve, along with significant changes in the political, social, and cultural world, other skills may need to be added to the list. I frame these as skills needed by general officers—indeed, the Pattons of the twenty-first century—but most of these skills and attributes should be important for every military leader at every echelon.

An Egalitarian Personality Trumps an Authoritarian Personality

In a widely cited book published in 1950, psychologist Theodore Adorno and several associates introduced the concept of the authoritarian personality.[10] Without delving into their theoretical interpretations, they described the authoritarian personality as being prone to rigid, right versus wrong thinking, a belief in strict adherence to rules and law, and the use of one's own authority to oppress and dominate others. The memory of World War II was still fresh in the mind of psychologists, who struggled to understand the Holocaust and other widespread inhumanity seen during that war.

A flurry of research into the authoritarian personality followed and lasted for over 30 years. Some believed that people with such personalities were attracted to certain professions, such as law enforcement or the military. Popular portrayals of military officers in films that appeared following the war often showed generals and other officers with personalities congruent with this stereotype. In a full-scale kinetic war like World War II, it was indeed often the case that killing massive numbers of enemy and destroying their capacity to wage war trumped subtle interpersonal skills, at least in the execution of specific missions.

A moment's reflection suggests, however, that a strong authoritarian personality was probably not optimal even in that time frame. The Supreme Allied Commander, five-star General Dwight Eisenhower, could not have begun to be effective in leading the force under his command, let alone working effectively with generals and statesmen from allied nations, if he used a strong authoritarian approach to his job.

Authoritarian methods may be expedient in getting a squad to attack a machine gun emplacement, but they are abject failures in building coalitions and highly functioning, interdependent teams.

I suggest here that generals of the twenty-first century, now more than ever, must have egalitarian personalities. Such personality types see all people as equal and work to influence (i.e., lead others) through establishing clear goals, effective communication, and displaying high social intelligence. Furthermore, given the broad spectrum of missions that even small military units will be given in future deployments, ranging from disaster relief to traditional war, even small unit leaders must show the ability to respect and listen to others, and only to rarely fall back on positional authority ("You will do this because I am your lieutenant and I said so!").

For platoon and other small unit leaders, this lesson has already been learned in Iraq and Afghanistan. Negotiation skills, empathy, and genuine concern for others get the job done when trying to get the local tribal chief to support a mission, where ordering him to comply may ultimately turn them to side with insurgents and cause the mission to fail.

It might be even more accurate to say that future military leaders should have personalities that allow them to switch leadership styles quickly. A company commander may lead a raid against an insurgent outpost in the morning and in the afternoon find himself being a mediator between competing tribal factions. Authoritarian leadership styles may occasionally be useful in the former condition, but the commander must be able to switch gears rapidly and tap into less authoritarian leadership methods when the circumstances dictate.

The Effective Leader will be Transformational, Not Transactional

In some ways this taps into a similar dimension of leadership as just discussed. The transactional leader is one who focuses on the costs and benefits of a given leadership situation. This leader focuses on solving current issues or crises, explicitly rewarding those who perform well, and punishing those who do not meet standards, and his or her effectiveness depends on the perceived ability to control and dispense rewards and punishment.

On the surface, the military seems to be the perfect setting for transactional leadership. There is a clear and rigid command structure, culturally approved rewards and punishments, and the commander has the legal authority to dispense these rewards and punishments as deemed necessary.

In contrast, a transformational leadership style focuses less on immediate objective outcomes (although they cannot be ignored), but more on understanding and responding to the needs, skills, emotions, and motivations of individual followers.

These leaders empower their followers. They provide the training and developmental experiences needed for followers to grow both personally and professionally. Such leaders build a climate of trust, cooperation, and respect for the overall mission among their subordinates.

Empirical studies of military leaders by psychologist Bruce Avolio and many others show that transformational leaders are more successful in most contexts than transactional leaders.[11] They are more respected, trusted, and liked than purely transactional leaders. Once again, transformational leadership is especially critical in missions with ambiguous goals such as nation-building. To the extent that future military deployments will be more characterized by such missions and less so by traditional kinetic fights, the emerging generals of the twenty-first century armed forces must appreciate and cultivate the transformational leadership style.

Leaders at All Levels Must Be Culturally Savvy

Americans in general seem to have a relatively low cultural IQ. Except for trips to Canada and Mexico, very few travel abroad. When they do, they often stay at western style hotels and visit tourist sites where the common language is English. It is no surprise, then, that American military personnel are largely culturally ignorant. The military recognizes this, and is making efforts to increase cultural sensitivity among its members. West Point cadets are now mandated to complete four semesters of a foreign language, instead of just two (interestingly, exceptions to this policy are granted to cadets who major in engineering), and they have many opportunities to travel and some even are able to spend a semester abroad.

A unifying theme in this book is that cultural literacy and an appreciation for other ways of viewing the world and doing things is vital for military leaders. Future military leaders must appreciate, acquire, and apply cultural knowledge if American armed forces are to be successful in future operations. Psychologists and other behavioral and social scientists must develop ways to increase this skill among military members. I suspect that taking additional language courses is not going to change core values much. The service academies may need to be more innovative in developing systematically designed multicultural experiences for their cadets, and include in their core curriculum courses in social psychology, sociology, and cultural anthropology.

Officers and senior NCOs are not the only ones who need this training. With modern technology, the lowest private's cultural blunder, caught on video and shared around the world, can cause strategic harm. Cultural ignorance, fear, and hostility contribute to the images we have seen of abused prisoners of war and desecration of enemy dead. This cultural education must begin in basic training for enlisted soldiers, precommissioning education and training for officer candidates, and continue through the career lifespan for enlisted personnel and officers.

There is a significant gap in our knowledge of how to increase cultural knowledge. Psychologists are faced with a daunting challenge of developing valid and effective ways of training and educating these skills in an increasingly diverse military. In turn, the military must recognize this training need as a strategic necessity and employ psychologists in the task of solving this problem. Not to do so may have drastically negative consequences for the military in the remainder of the twenty-first century.

Adaptive Thinking is Crucial in a Diverse and Volatile World

A friend of mine, a former army infantry commander, said that given the stunning power of modern weapons, tactical decisions are relatively easy. "Right, left, or up the middle" is how he put it, meaning that with modern firepower, once the objective is identified, the question of how to attack the objective is a relatively easy decision. He went on to say that a moderately good but quick decision is better than a perfect but slow decision. These comments are based on the application of kinetic force to targets. My friend went on to acknowledge that the hardest things he did while leading soldiers in combat in Iraq involved tasks that included negotiations with local leaders, serving as a mediator in disputes, building coalitions among locals, and a myriad of other tasks that required things that the army did not really train him to do.

Adaptive thinking refers to the ability to rapidly assess and reassess situations, and make changes in plans quickly, while on the fly.[12] While this certainly applies to traditional war fighting skills, it may be much more critical in the other situations that the military finds itself in today and is likely to encounter in the future. There is a growing cognitive science to adaptive thinking, but this needs to be extended to the decision-making tasks and conditions typical of contemporary and future military operations. Is adaptive thinking an innate ability, or something that can be trained? Is it task-specific, or are adaptive thinkers good at diverse tasks? Can valid and reliable tests be developed to help the military identify and then groom its most adaptive thinkers? Psychologists can help provide answers to these questions, as ensuring that future military leaders are skilled in this domain is essential to future success.

Leaders Must Be Technologically Smart

Human factors psychologists have long studied the role of the human operator in complex systems, but they have not paid much empirical attention to learning how leadership and technology interact to affect organizational performance. The future general does not need to personally know how to operate each technological system under his command—but he does need to know the collective capability of these systems and how to leverage them to succeed in different types of missions. More importantly, he must remember and appreciate that humans are the most critical component of even the most highly sophisticated technological system. And, as discussed in Chapter 9, the future general must have a proficient working knowledge of the command, communication, and control technologies at his disposal.

Future Military Leaders Must Be Politically Well-Informed

This does not apply only to generals. Lieutenants and other small unit leaders must know the political geography of the areas in and around where they are deployed. Political and cultural knowledge overlap and both are crucial to good leadership. Particularly at strategic levels, leaders must be sensitive to the political realities of the allied nations involved in the operation.

Social Intelligence Is a Core Job Requirement

This is a bit different from cultural intelligence. The latter focuses more on the customs and beliefs of those countries or cultures in which the military is deployed. Understanding social customs and courtesies common to Islamic nations is an example. Rather, social intelligence is the ability to recognize and appreciate the thoughts, emotions, and motivations of one's colleagues, peers, subordinates, and superiors.

The military has never formally assessed social intelligence in its members. Highly successful officers and NCOs are often high in this trait, but I suspect this is due to social Darwinism, in the sense that individuals who are lacking in these skills fail to be promoted to higher ranks. Given the social complexities of the current and future military, psychologists can play an important role in assessing and developing social intelligence through formal programs that may be implemented in various military educational and training venues.

A Diverse Military Requires Diverse Leaders

As discussed in Chapter 7, the military is becoming increasingly diverse. A leader who does not tolerate or accept religious practices other than his own, who is misogynistic, or who is homophobic will fail abjectly. Moreover, the military must strive to promote a more diverse set of leaders to its highest ranks. A diverse military led by predominantly white, heterosexual Christian males cannot effectively lead or manage the diverse military of the future. The military recognizes this, but more can be done. As the technology, machinery, and missions of the military change, we may expect to see (at least from the military's perspective) an even more diverse force. People who are overweight, physically challenged, and older may all come to offer important skills to the military. Psychology has a clear role in helping the military develop better and more effective diversity programs.

The Rules of Engagement Have Changed: Do Not Kill or Be Killed

In the movie *Patton,* General Patton gives a speech to his troops in which he claims that there is no glory in dying for one's own country, but there is considerable glory in making the enemy soldier die for *his* country. The world is becoming increasingly averse to systematic killing seen in war. Unlike wars of the previous century, images of dead soldiers and civilian noncombatants can be instantly broadcast across the globe by anyone with a smart phone.

In this environment, the mindset of the future military leader must be ever more focused on attaining the objective with minimal loss of life. This is especially true when combating foes who represent ideological groups rather than nations. This represents a psychological orientation toward killing that is almost the antithesis of previous wars, where a high body count signaled operational success. Future generals must be able to kill in large numbers where necessary, but must also understand that many of the missions they will face will require a more nonlethal approach. The ability to switch between a killing and nonkilling schema is another instance of the adaptive thinking that will characterize effective military leaders of the future.

The Effective Military Leader Will Be Skilled in Working with Nonmilitary Agencies

In my experience, many officers do not like working with civilians, especially those from other agencies, both government and nongovernment (NGO). In future deployments, the military will work ever closer with the Department of State, Homeland Security, the Federal Bureau of Investigation, and the Central Intelligence Agency, just to name a few. Moreover, they will also need to work effectively with organizations like the United Nations Children's Fund (UNICEF), the Red Cross, and many other disaster relief and aid organizations.

Tied closely to political skills and social intelligence, the ability to work with these diverse organizations will be a necessary ability for future military leaders. This represents a significant change in the psychology of the military leader, and psychologists and other social scientists again must help the military develop ways to assess and develop these skills in its emerging leaders.

CONCLUSIONS

I have often wondered how General Patton would have fared had he not died in a tragic motor vehicle accident following the conclusion of World War II. What if he had lived and become the Army's Chief of Staff? Would his hard as nails, authoritarian leadership style (at least as depicted in the media) have been as effective in managing a large peacetime army as they were in marching an army across Europe to defeat the Germans?

It is hard to picture the twenty-first century Patton building coalitions, negotiating patiently with warring tribes, or cooperating with nongovernment agencies to achieve nation-building. I doubt he would be very tolerant or understanding of soldiers with posttraumatic stress disorder. Don't Ask, Don't Tell would be unthinkable to him.

As we have discussed, military missions of the twenty-first century may begin with a period of traditional kinetic warfare but, like Iraq in 2003, will quickly evolve into peace keeping and nation-building, something that the military refers to as operations other than war (OOTW). The twenty-first-century Patton must bring his social and cultural skills to bear on the OOTW tasks in order to achieve the nation's larger strategic goals.

The leader attributes I have identified require a combination of talent and skill. Some of these attributes may be trait-like and thus require screening and selection, while others may be more amendable to training. It is a lot to ask to find all of these attributes manifested in a single individual. But the military grows its own strategic leaders and has over 30 years to identify, train, develop, and nurture those who will lead and command in 2050 and beyond. It has the advantage of an organizational system designed for this job. Moreover, at the highest level of strategic leaders, the military does not have to find or develop very many. The army, with a force of approximately 500,000 soldiers, has only 12 four-star generals. It and the other branches of the military can start today to systematically identify and develop the Pattons of 2050. And psychologists must work hand in hand with the military to help it in this task.

NOTES

1. From "War Is A Racket," a speech delivered in 1933, by Major General Smedley Butler, USMC; see http://quotes.liberty-tree.ca/quotes_by/major+general+smedley+darlington+butler.
2. US Department of the Army, *Army Leadership*.
3. Christopher Peterson and Martin E. P. Seligman, *Character Strengths and Virtues*.
4. Michael D. Matthews et al., "Character Strengths and Virtues of Developing Military Leaders," S57–S68.
5. Michael D. Matthews, "Character Strengths and Post-Adversity Growth in Combat Leaders."
6. Patrick J. Sweeney, "Trust," 252–77.

7. The origin of this model is described by Patrick J. Sweeney et al. in "Trust," 163–81; Figure 10.3 represents a more fully developed version of that model.

8. Michael D. Matthews, "Where Eagles Soar."

9. Matthews, "Character Strengths and Post-Adversity Growth in Combat Leaders."

10. Theodor W. Adorno et al., *The Authoritarian Personality.*

11. For a good introduction to the concept of authentic leadership as it applies in military settings, see Sean T. Hannah, "The Authentic High-Impact Leader," 88–106.

12. Much of the research on adaptive thinking and leadership in military contexts is conducted by Dr. Stephen Zaccaro of George Mason University. For a good introduction to his work, you may refer to Zaccaro et al., *Leader and Team Adaptation: The Influence and Development of Key Attributes and Processes.*

11

BUILDING BETTER SOLDIERS
THROUGH SCIENCE

I know not with what weapons World War III will be fought, but World War IV will be fought with sticks and stones.
Albert Einstein[1]

Albert Einstein lived to see the horror of nuclear war. Physics—largely based on Einstein's work—radically affected the outcome of World War II. No one can look at the pictures of Hiroshima and Nagasaki and not feel regret that such weapons had to be employed to bring a speedier conclusion to the war than would have been possible through the use of conventional weapons and a land invasion of Japan. The fact that such weapons have not again been used in nearly 70 years speaks to the utter futility of nuclear war. There would be little left to dominate after such a war.

Of course the advent of ever more powerful nuclear weapons did not end war. It did change its nature, with smaller and more regional wars such as the Korean and Vietnam conflicts becoming the norm. These and the more recent wars have thus depended once again on conventional forces, rather than the power of the atom, to achieve political and military aims. This, in turn, places all the more reliance on the capabilities of soldiers to ensure a favorable outcome.

The military turns to science to enhance its capabilities to wage war. Major General Robert Scales, US Army (Ret.), as discussed earlier in this book, points out that different sciences have provided an edge for each war since the beginning of the twentieth century. Commenting on a taxonomy of war first introduced by Alan Beyerchen, Scales points out that chemistry was the key science in World War I. Physics decided the outcome of World War II. The Cold War was heavily influenced by information technology and computer science. Scales maintains that the current wars are dependent on psychological science.[2]

What can we expect in the coming years? To be sure, physical scientists and engineers are hard at work developing more potent weapons systems. These might include high-energy beam weapons, rail guns that fire huge numbers of projectiles at supersonic speeds using electricity rather than gun powder, and robotic weapons that operate in and under the sea, on the ground, in the atmosphere, and in space. Weapons will continue to get smarter and more precise. Advances in intelligence monitoring technologies and information technology will allow almost instant military responses to military threats.

Whatever advanced systems the physical scientists and engineers deliver, their use will be dictated by human beings. Politicians will decide where and with whom to engage in war. Strategic military leaders will decide how to utilize military force to achieve the aims dictated by the politicians, and soldiers will do the hard work of fighting and winning wars. War will continue to be a fundamental human endeavor, as it always has been. And psychology, as the major science that systematically studies human behavior, will continue to be a necessary part of preparing for and engaging in future war.

A comment on the ethics of science and war is in order. Einstein, despite being morally opposed to war, came to see it as a necessity as he became aware of the events unfolding in Europe in the 1930s.[3] As educated people, scientists have always found war to be repugnant, but they have also realized that a strong military is necessary to preserve a greater good. Psychologists are no different. From World War I until the present, psychologists have lent their expertise to improving the combat capability of the military. The previous chapters of this book have looked at how psychologists have improved training, selection, human–machine interface, and soldier resilience, just to mention a few of their contributions.

In this chapter I want to invoke a bit of imagination and speculate on ways that psychologists might influence how the military prepares for and fights wars of the future. Writers of fiction are granted poetic license, or a bit of leeway in interpreting the world, as they endeavor to depict an idea. I ask the reader to indulge me in something akin to poetic license as I ruminate on the role of psychology in future war.

That said, my suggestions will not be science fiction. Noted experimental psychologist F. J. McGuigan discussed which topics are and are not within the purview of science.[4] Questions that can be assessed through empirical means with tools that are currently available are most appropriate to all science, including psychology. Other questions exist that are potentially answerable through empirical methods, but the tools are not yet available. Finally, some questions simply are not addressable using the scientific method. "Is there a god?" represents such a question, whereas "Does belief in a god improve mental health?" is a valid question for psychologists to address.

Psychological science will help build better soldiers in the future. In the subsequent pages, I will follow McGuigan's guidance and focus on topics that are presently subject to empirical inquiry, or are likely to be so within the coming years. The military is currently funding research on some of these potential applications. In some cases I am extrapolating from current science to what may plausibly be possible in the near-time future. Many of these suggestions cross traditional boundaries between psychology, biology, engineering, and other sciences. Collectively, however, they will advance the skill, strength, and resilience of soldiers of the future.

ENGINEERING THE RESILIENT SOLDIER

War inevitably takes a toll on soldiers, both physically and psychologically. With enough combat trauma even the best soldiers will eventually become combat ineffective, and many will suffer behavioral and psychological damage.

In the past, the military paid scant attention to preparing soldiers for the psychological and emotional realities of combat. It was assumed that traditional military training, with its emphasis on competence, cohesion, and morale, would protect the soldier from combat stress. There are indeed aspects of traditional military training and the social dynamics of being in the military that do help soldiers respond adaptively to stress, but we are learning that this alone is not sufficient. More systematic efforts are needed.

Historical evidence and contemporary experience show that as many as fifteen percent of soldiers may fall victim to clinically diagnosable combat-related stress disorders.[5] An unknown number experience subclinical symptoms and cope silently, or self-medicate with alcohol and drug abuse. The military's response through the years has been to invoke a disease-model approach: That is, they do little formally to promote mental health, and respond reactively with psychological and medical treatments for soldiers who do ultimately present with significant pathologies.

In response to historically high rates of suicide, PTSD, and a host of other behavioral and psychological disorders, the military has recently begun to adopt a proactive, wellness-based approach that aims to imbue soldiers with the cognitive, social, and emotional skills to respond in a resilient manner in the face of combat stress and deployment adversity. Programs like the Army's Comprehensive Soldier Fitness (CSF) exemplify these attempts to prepare soldiers for the psychological consequences of combat *before* they deploy, rather than simply treating psychological casualties *after* they occur.

Programs like CSF represent a seismic shift in the military's approach to preparing soldiers for combat. In particular, CSF—which directly impacts over 500,000 active duty army soldiers—represents what may be the largest single intervention based on empirically derived psychological principles. It is too early to fully evaluate the success of CSF. But perhaps the most important outcome is the change of attitude of the military toward the role of psychology, from a clinical-based treatment model to a wellness-based training model.[6] This greatly expands the role of psychology in the military, as well as exponentially increases its potential impact on individual soldiers, their families, and army civilians.

It is interesting to speculate on innovative ways the military may turn to in further increasing soldier resilience. This is critically important to the military, particularly in an era when military forces are small (compared to the large standing forces of the twentieth century) and the contributions of every soldier matters. If the military could develop an immunization that would prevent combat stress and its associated disorders, then the combat effectiveness of the force would be increased substantially. This increased combat effectiveness could be a difference maker, especially in prolonged wars such the Afghanistan War.

The greatest advances in resilience building in coming years may come from behavioral neuroscience. These scientists are developing an ever more sophisticated understanding of the human brain. The brain responds in complex ways to stress. The stress response includes anatomical brain regions, including the prefrontal cortex and the limbic system; brain interactions with the endocrine system that mediate the release and regulation of stress hormones; and the subtle role of a host of neurochemicals that modulate brain response to stress and anxiety. The Defense Advanced Research Projects Agency (DARPA) is reportedly funding research that shows promise of developing a drug that would, in essence, erase traumatic memories.[7]

I foresee, therefore, advances in brain science that may one day result in drugs that are capable of regulating the brain's response to combat stress. These drugs may alter the synthesis, release, or deactivation of brain chemicals (known as neurotransmitters) known to be related to the stress response, such as neuropeptide Y or norepinephrine. Brain chemistry is exceedingly complex and undesirable side effects are common (listen carefully to the disclaimers given on television ads for drugs used to treat depression, for example). In theory, though, it may be possible to develop drugs that have very specific effects on these brain chemicals, and in doing so mute the pathologic response many soldiers might otherwise experience.

The human stress response, while mediated by the brain, also involves an endocrine system response including the release of hormones from the adrenal gland, located on the kidneys. This hormonal response to stress is mediated by the HPA axis, which involves the hypothalamus (a small but critical area near the base of the brain that mediates basic motivations), the pituitary gland (aka the "master gland," which releases a host of hormones into the bloodstream that affect a variety of target organs), and the adrenal cortex. Activation of the HPA axis causes the adrenal cortex to release the hormone cortisol, which increases metabolic activity and prepares the person for emergency situations.

The human stress response also involves the activation of the sympathetic branch of the autonomic nervous system. The autonomic nervous system regulates smooth muscles and glands and the heart. Activation of its sympathetic branch results in the fight-or-flight response. It increases heart rate, directs blood away from the digestive system and into the muscles and brain, and quickly prepares the body for an emergency response. After the threat is removed, the parasympathetic branch of the autonomic nervous system slowly restores homeostasis.[8]

Both the HPA and sympathetic responses to threat are normally adaptive. It is good to quickly have more energy when faced with danger. However, when people find themselves in prolonged situations—and a combat deployment is a perfect example—where they are continually cycling between activation and deactivation of the stress response, this may have deleterious impacts on the brain and other body organs. Moreover, people differ a good deal in their response to stress. This reactivity is likely modulated by how the person frames events and coping skills they have learned—things that programs like CSF teach soldiers.

Changing the HPA and sympathetic responses of soldiers to combat stress, therefore, may also be a strategy for further enhancing soldier resilience. As I mentioned, this may come through more effective training procedures and psychologically based interventions, but pharmacologically based approaches are also possible.

The emerging field of epigenetics also holds promise for improving soldier resilience. Epigenetics looks at how the environment alters the expression of genes. This modification of gene function occurs without changing DNA sequences. Stress may alter basic biochemical processes—such as methylation—that act to turn genes on or off. It may prove to be the case that soldiers who are predisposed to combat stress disorder have a different genetic profile than resilient soldiers. It is theoretically possible to turn off a stress reaction, or, alternatively, to turn on a resilient reaction by devising ways to modify the expression of genes. Stay tuned: This may be one of the more fruitful applications of modern genetic science not just to the military, but to people in all walks of life.

At the farther reaches of plausibility, it is at least theoretically possible to change brain tissue. The field of brain machine interface (BMI) looks at ways that technologies may be wired to directly impact the brain.[9] As the specific architecture of the brain anatomy that is involved in the human stress response is discovered, temporary or permanent alterations of those parts may be designed to result in a resilient response to combat stress. Stress is known to affect the hippocampus, a forebrain structure involved in certain types of memory. Under severe and prolonged stress, the hippocampus may reduce its volume. Strategies might be discovered that could prevent this response to combat stress, thus making a more resilient response more likely.

The second and third order effects of preventing emotional trauma among soldiers would be profound. Reducing or eliminating stress-related combat injuries would be equivalent to adding an additional division of soldiers to the fight, vastly increasing combat power. When soldiers return from war personal, family, and social problems would be greatly diminished. Moreover, the treatment of psychological casualties of war is immensely expensive. Reducing lifespan health costs by substantially reducing the incidence of stress-related disorders would yield billions of dollars of savings in the veteran's health care system.

There are many caveats in this application of psychology and neuroscience. Any methods used to alter brain or endocrine system functioning must be totally reversible. These neural systems are critical to normal, day-to-day living. Additionally, managing the brain's emotional and stress systems would likely affect the soldier's cognitive reactions to combat. The link between cognition and emotion is a strong one, and it is unclear how reducing emotional responsiveness to combat stress might affect how the soldier thinks about and interprets his or her combat experiences. Soldiers must not lose their aversion to killing as they return to a peacetime military and ultimately to civilian life. It is essential that resilience programs not result in a person who is insensitive to the value of life.

In any event, psychologists, through designing improved methods of regulating psychological reactions to combat, and behavioral neuroscientists with their advances in understanding the biology of combat stress, will develop radically new ways of preparing soldiers for the emotional realities of war prior to deployment, and for returning to a normal and peaceful life after combat. These changes are coming, and it is important to both the soldiers and to society to get this done right.

ENGINEERING THE 24/7 SOLDIER

One of the first casualties of combat operations is sleep. During intense combat, soldiers must learn to function with little or no sleep for many hours and sometimes days at a time. Soldiers in command and control centers work continuously and catch sleep when they can, as shown in Figure 11.1.

Soldier performance is closely tied to sleep. I conducted research in Norway where I gave Norwegian Army Academy cadets a battery of four computer-administered cognitive tasks at several points during a prolonged field exercise. During this exercise the cadets operated with no continuous sleep for eight days, with the exception of a four-hour sleep period mercifully given to them halfway through the exercise. They might have managed very short naps or microsleep during the remainder of the period, but for the most part were engaged in physically demanding military missions day and night throughout the period. We found that, compared to baseline measures taken prior to the exercise, their cognitive performance began to plummet after about 36 hours without sleep. It continued to deteriorate through the remainder of the exercise and took up to 48 hours to recover following the end of the operation.[10]

In another study with Norwegian cadets during a similar exercise, I found that when asked to self-rate their performance during different missions that make up the

FIGURE 11.1 Long and unpredictable workdays, coupled with the adversity and stress of combat deployment, make it difficult for soldiers to remain vigilant at all times. (Photo courtesy of David Uthlaut.)

exercise, they became progressively less able to do so. Compared to expert observers who also rated their performance, the cadets' own ratings showed virtually zero correlation. In short, they lost the ability to self-monitor.[11]

The impacts of sleep deprivation are even greater for leaders who must make important decisions during combat than for soldiers who may be tasked with relatively simple jobs. For example, the decision-making performance of tank commanders suffers substantially during conditions of sleep deprivation, and may lead to a variety of potentially grave errors during combat operations.[12]

How might we create soldiers who could fight effectively for 100 hours or more? Like resilience, the solution probably resides in changing brain functioning. Also like resilience, this depends both on certain anatomical features of the brain and on neurochemicals associated with sleep and waking cycle. For example, a structure called the reticular formation extends from the medulla oblongata, located in the hindbrain, up into the forebrain. Damaging this structure decreases cortical arousal and stimulating it increases activity in the cortex. A portion of the reticular formation, the pontomesencephalon, is responsive to sensory input and drives the release of neurochemicals associated with wakeful alertness. The locus coeruleus, also part of the reticular formation, is inactive during sleep but is reactive when meaningful stimulation, such as hearing your named called, is received. Stimulation of this structure encodes memories and also increases wakefulness.

A host of neurochemicals and hormones impact sleep. Various brain structures release neurotransmitters that, in turn, affect behavior. For example, cells in the midbrain release acetylcholine and glutamate, which increases brain arousal. Evolutionarily old parts of the forebrain release neurotransmitters that both excite (acetylcholine) and depress (gamma-amino-butyric acid or GABA) brain activity. The hypothalamus, vital to modulating basic motivated behaviors including eating, drinking, sex, and fear responses, releases both histamine and orexin. Histamine increases arousal while orexin appears to *maintain* wakefulness. Many other brain regions and associated neurotransmitters regulate sleep and all are candidates for research that may lead to revolutionary drugs designed to induce and maintain arousal and wakefulness.[13]

A small structure deep in the brain called the pineal gland helps regulate the body's normal 24-hour sleep/wakefulness cycle. Sunlight strikes the eyes and affects the hypothalamus, which, in conjunction with the pineal gland, regulates the release of melatonin, a hormone found in relatively low concentrations during the day that increases about two hours before bedtime.

Physiological psychologists and other neuroscientists may develop ways of manipulating these sleep systems in ways that greatly delay the body's need to sleep. Given adequate nutrition, soldiers could conceivably fight effectively for several days at a time given such interventions. It is unlikely that methods could be developed that permanently eliminate the need for sleep, but extending wakefulness for periods up to, say, 100 hours would have a huge impact on combat force and effectiveness.

Like strategies to increase resilience, extending the soldier's ability to fight by temporarily eliminating the need for sleep would also have many secondary effects. One small unit could do the work of two or more units, because its soldiers would not need to rotate shifts. Additionally, soldiers and their leaders who are alert would make fewer decision errors. Fratricide would be reduced. Better tactical decisions would be made. Soldiers could use their full cognitive resources in seeking and identifying enemy activity, executing the tactical fight, and dealing with the aftermath of a tactical engagement. Pilots could fly continuous operations. Radar and sonar technicians would reduce their error rate. In short, every aspect of military performance could be enhanced.

A cautious approach is again needed. Experience has shown that manipulating a given aspect of brain functioning may have significant and unexpected ripple effects. It is not worth harming soldiers in the long run to achieve short-term tactical advantages. Psychologists are well poised to assist the military in devising ways of diminishing the need for sleep and evaluating both the benefits and potential costs to individual soldiers and their units.

BUILDING KILLER SOLDIERS

Some years ago I was talking with General Wayne Downing, US Army (Ret.). General Downing was a highly regarded officer who served in wars from Vietnam to

Afghanistan. He was most noted for his command of Special Operations Command, including Delta Force. He made a comment that I will never forget. He said that in his experience the majority of soldiers were not effective killers. Specifically, he said that in a typical infantry platoon of about 30 soldiers, he was lucky to have three or four who would effectively kill the enemy. The rest, he said, ended up basically being "ammunition carriers."

This statement raises lots of interesting questions about both the selection and training of soldiers, but also about human nature itself. He went on to say that, even in special forces, many soldiers—however good they were at various military skills— really were not effective killers. So despite years of training, some and perhaps most soldiers are not effective in the basic task of killing others.

General Downing is not the first to make this observation. David Grossman reviewed reports from wars from the nineteenth and twentieth centuries, which indicated that most soldiers would not fire their weapons actively, if they fired them at all.[14] Brigadier General S.L.A. Marshall estimated that during World War II perhaps 15 to 20 percent of soldiers would effectively fire their weapons. Many would engage in what Grossman calls "posturing," which involves holding the rifle in a threatening manner but firing into the air or at an angle that would greatly reduce the possibility of killing an enemy soldier.

Increasing the number of killers in an infantry platoon would greatly increase combat effectiveness. This is especially true in the current era, where for economic and political reasons the nation maintains a small standing military force. If the percentage of killers could be doubled, the impact on mission effectiveness—at least for those occasions where the military is employed in traditional combat operations—would be greatly enhanced.

The psychology of killing is not well understood, particularly when trying to understand, predict, and control socially acceptable killing. Psychologists usually view killing as a pathology. There have been volumes written about the psychology of serial killers, for instance, but very little research has been reported on what I might call "adaptive killing." Adaptive killing includes legally and socially mandated authority to kill others in the name of a greater cause. In addition to the military, there are other institutions that allow adaptive killing in our society. Law enforcement officers can kill in the line of duty. State officials can execute those convicted of capital offenses. Each citizen has the right to use lethal force under very well defined and limited circumstances.

Most of the literature on adaptive killing is anecdotal. David Grossman's book *On Killing* is a prime example. In it he reviews the history of killing in combat and then discusses the social and psychological components of killing. It is not surprising that we are left with only correlational or anecdotal accounts of adaptive killing. It would be unethical to conduct true experiments in this extreme aspect of human aggression.

Current and future psychological science may, however, offer ways to increase adaptive killing among soldiers. The following are a few avenues that psychologists might employ toward this end.

Adaptive Killing as Behavior: An Operant Conditioning Approach

From a behavioral perspective, organisms can be shaped to perform almost any behavior that is consistent with their biological constraints through the systematic use of reinforcement contingencies. Known as operant conditioning, the basic approach involves rewarding a person for showing the desired behavior and ignoring or even punishing incorrect responses. In the military context one could define the target behavior as firing a rifle at the enemy using proper firing procedures (breath control, sight picture, trigger pull). Then, in training, soldiers could be shaped as their firing behavior increasingly approached the goal behavior. Successful firing would then be reinforced through praise, special privileges, and the like.

An important component to this approach would be to more closely simulate real combat during training. The military has long been able to train soldiers to fire with high accuracy at paper or metal targets. The question for the psychologist, as well as the military trainer, is how does this firing generalize to shooting at real enemy soldiers. Technology offers a way to increase the external validity of marksmanship training. After basic rifle marksmanship training in the field, extensive training in immersive combat simulations would allow trainers to condition the soldier to fire at very realistic human targets. In highly sophisticated simulations the successful killing of enemy soldiers could include visual, auditory, tactile, and olfactory feedback. Properly designed, these simulations could not only train soldiers to kill more effectively, but also help trainers sort out those who are not capable of displaying lethal aggression toward others.

An advantage to this approach is that trainers could also use discrimination training procedures to teach the soldier what constitutes an appropriate human target. In this context, discrimination training would involve rewarding the soldier for using deadly force on an appropriate enemy target, but withholding reinforcement or perhaps giving punishment for using deadly force on an inappropriate target, such as a civilian noncombatant.

Adaptive Killing and the Brain: Biological Approaches

Human aggression is linked to a variety of genetic, hormonal, and central nervous system factors. A significant problem to leveraging emerging knowledge of the psychobiology of killing is the ability to increase adaptive killing without increasing general levels of aggression. It may be desirable to increase adaptive killing among soldiers, but only if an enhanced propensity to kill can be constrained to the battlefield. Even in combat, when the battle ends, soldiers must immediately cease in their use of deadly force. In general, biologically based approaches would likely be harder to control than a training/learning-based approach.

For instance, it might be possible to identify and manipulate genetic processes involved in aggression. Advances in genetics might allow scientists to identify genes that are instrumental in expressing aggressive behavior. Epigenetic interventions could conceivably be developed to stimulate or inhibit the activity of these genes. Before deployment, soldiers could be given drugs that affect the methylation of targeted genes, allowing them to be more aggressive during combat missions. However, this would be of little use if the intervention could not be readily applied and withdrawn. Even long combat deployments involve relatively few periods of actual combat. It would not be adaptive to have large numbers of soldiers genetically prepared to kill in the long intervals between engagements.

In another context we discussed brain machine interface. As the brain structures involved in lethal aggression are better identified, it is theoretically possible that their function could be modulated through brain stimulation. For instance, structures in the limbic system—the part of the brain that helps regulate emotional expression— could be activated by remotely controlled microelectrodes at the initiation of combat. If the role of the limbic system and its relation to other brain structures is properly understood, then such stimulation could set the occasion for increased adaptive killing. Better yet, when the fight ends, these systems could be turned off.

There are a host of other biological factors related to aggression, but they would be difficult to regulate. Diets low in tryptophan, an essential amino acid, are associated with less turnover of the neurotransmitter serotonin in the brain. This in turn has been linked to human aggressive behavior including violent crimes and violent suicides. Prenatal exposure to central nervous system stimulants, even legal ones like nicotine, has been linked to violent crime many years following birth. Subtle irregularities in other brain chemicals are also associated with aggression.[15]

Cognitive neuroscientists might help the military explore ways of manipulating these and other biological processes to increase adaptive killing in soldiers. I should be clear in stating that this is not presently attainable, but the potential for these sorts of approaches exists as science provides an ever more comprehensive understanding of the brain.

Adaptive Killing: Cognitive-Based Approaches

How a person frames and interprets situations affects all behavior including adaptive killing. To become adaptive killers, soldiers must learn to modify schemas they have acquired previously in their lives that prohibit killing. Learned through social interactions with parents and community, and reinforced through religious and educational institutions, a recruit comes to the military with strong beliefs about the inappropriateness of taking human lives.

Traditional military training attempts to change this belief about killing so that the recruit can come to terms with the possibility of killing other human beings. Cognitive and clinical psychologists can leverage existing methods for effecting cognitive change to achieve this effect. Cognitive-based therapy techniques, which focus on eliminating irrational thoughts and beliefs, could be focused on changing a soldier's belief structure regarding killing. These interventions could be integrated into immersive simulations to promote the conviction that adaptive killing is permissible. Moreover, these same techniques could be employed to teach soldiers not to have guilt or excessive remorse over the killing of enemy soldiers, and at the same time train soldiers to discriminate between adaptive killing and inappropriate use of lethal force.

Some might argue that such training is unethical. I would argue that as long as our political leaders and the world situation results in the need to employ soldiers to kill the enemy, it is unethical *not* to employ psychological science to help them do so more effectively, both in terms of accomplishing the mission and in responding in an adaptive psychological and emotional manner following their service.

Adaptive Killing and Returning to Civilian Life

Training soldiers to kill has always been a task fraught with practical and ethical considerations. Of equal importance to training soldiers in adaptive killing, psychologists must help the military develop ways to untrain this ability before returning soldiers to civilian life. The same approaches that may be used to increase adaptive killing can be used to decrease it as well. Adaptive killing behavior can be extinguished in simulations. Brain machine interface, epigenetic, or other biologically based interventions may be withdrawn. Cognitive therapy can be delivered to restore the normal abhorrence to killing.

In twenty-first-century war, it is not enough to create proportionally more killers among the military's ranks. The nature of modern warfare requires soldiers to quickly shift from relatively peaceful operations to full-on combat and then back to peaceful operations in short periods of time. Psychologists must leverage behavioral, biological, and cognitive strategies to enable tomorrow's soldiers to make these transitions quickly and effectively.

ADDITIONAL ADVANCES IN SOLDIER SCIENCE

Enhancing Strength

The idea of using exoskeletons to increase strength and stamina is not new, but thus far efforts to develop useable systems have not been successful. With advances in brain machine interface, however, it may be possible to design an exoskeleton that seamlessly integrates with human movement and brain activity. In essence, the exoskeleton may become an extension of a soldier's physical body. Thus equipped, an infantry soldier could carry much heavier combat loads, be able to move or lift heavy objects, and sustain activity for long periods of time with less fatigue on his own neuromuscular system.

Psychologists will play a critical role in the design of such systems. Human factors specialists must ensure that the human and the system are well integrated in order to maximize performance. Cognitive neuroscientists will help integrate the system with the brain. Organizational psychologists must explore how greatly enhancing soldier strength and stamina could impact the size and structure of units that employ this technology.

And how will a soldier react to these increased physical capabilities? Could having superhuman abilities cause some soldiers to misuse their enhanced strength? Like other soldier systems, training would need to include instruction and indoctrination of when and how to employ it. But a system that is truly an extension of one's own central nervous and muscular systems might give some soldiers a distorted sense of power with the resultant potential to do harm to others. Clinical psychologists should be involved in the design of these systems and help develop training procedures to help soldiers adapt to the psychological impact of their newly imbued strength.

Advanced Prosthetics

General Eric Shinseki (Ret.), former US Army Chief of Staff, was severely wounded in Vietnam. His wounds included the partial loss of a foot. In that era, an amputation was almost always a million-dollar wound resulting in a medical retirement from active duty. The prosthetics of the time were good enough to allow the amputee to walk, but were far less advanced than current ones. General Shinseki successfully petitioned the army to remain on active duty. He did not ask for any special consideration, and for the next 30-plus years he took the army's twice-yearly physical fitness test (the APFT) that includes a two-mile run that must be completed in a specific age-adjusted time. Never one to complain, Shinseki successfully completed the run component of the test each time, a painful and true test of his devotion to the army and its mission.

Today's prosthetic devices are light years ahead of those from just a generation ago. Each year I participate in the Achilles Hope and Possibility Five-Miler, a run designed to bring hope, inspiration, and the joys of achievement to people with disabilities or war injuries. Organized by Trisha Meili, who was known to the world as "the Central Park jogger" after being brutally raped in New York's Central Park in 1989, the event includes many wounded warriors celebrating their ability to overcome their injuries.[16]

It is a humbling and inspiring experience to run with (and often be passed by) men and women running with prosthetic legs, or speeding along on hand-cranked bikes operated by those whose injuries are too severe to allow a prosthetic leg. You have all seen these prosthetic legs—South African Olympian Oscar Pistorius, a double amputee, competed in the 2012 London Summer Games in several running events, using his high-tech running blades.

As good as these prostheses are, the technology is coming that will allow the development of artificial limbs that link directly to the central nervous system. At West Point we recently assisted in pioneering research with a bionic foot. The bionic foot—which greatly resembles the real human foot—is equipped with sensors that detect the activity of nerves located in the leg just above the amputation, and feed that information to a small wearable computer that allows the foot to behave mechanically during running and walking like a real foot. In our engineering psychology lab, amputees used this foot, shown in Figure 11.2, to walk and run on a treadmill. Based on what we learned in the lab, amputees were later able to use the bionic foot to run outside on West Point's 400-meter track.

Imagine if this technology existed when General Shinseki was wounded! How many other soldiers with similar or even more severe wounds might have been able to remain on active duty?

Even greater advances are on the horizon. With brain machine interface, next-generation prosthetics will be wired directly to the brain, not just to peripheral

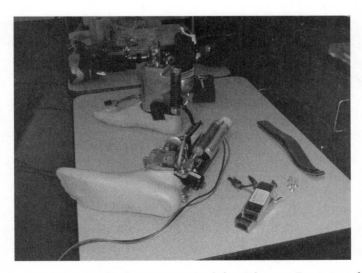

FIGURE 11.2 A bionic foot, tested in the Engineering Psychology Laboratory, Department of Behavioral Sciences and Leadership, US Military Academy. (Photo courtesy of the Engineering Psychology Lab.)

nerves. Here is where it really gets interesting. Not only should it be possible to wire the motor centers of the brain to a prosthetic limb, it should be possible to wire the prosthetic to the sensory centers in the brain. This dual capability would allow the amputee not only to *move* the foot, leg, hand, or arm as if it were his or her real appendage or limb, but it would allow the user to *feel* the feedback from the prosthetic. In short, such prosthetics would move and feel like the real limb.

The impact on the military is obvious. Modern body armor protects the central core of the body and the head pretty well, but blasts from IEDs and other explosive devices often tear apart the limbs. These injuries can be psychologically devastating, not to mention the impact they have on the organization in terms of losing the future contributions of highly trained and motivated soldiers. The introduction of these next-generation prosthetics will thus have profound personal and organizational effects.

With these prosthetics, many soldiers who would currently be forced into medical disability may be presented with the opportunity to remain in the active duty military. When they leave the military—either remaining until retirement or earlier—they will be far less constrained in vocational options than is currently the case.

The psychological implications of these prosthetics are obvious. The ability to retain full or nearly full mobility and to engage in important and meaningful work will greatly improve the postinjury adjustment of affected soldiers. And knowing that these technologies exist may help ease precombat fear that soldiers experience in anticipation of losing a limb. Finally, the military will retain the services of members who in past wars would have been relegated to the sidelines because of their injuries.

These advancements in prosthetics are coming soon, and are not a mere abstraction. I have known several West Point graduates who have suffered horrific injuries in combat. Just recently, I learned that a 2006 graduate of our Engineering Psychology Program lost both legs above the knee, one arm, and three fingers from the remaining arm in wounds suffered in combat in Afghanistan. These advances cannot come soon enough. Our brave soldiers deserve the best prosthetics science has to offer.

Robotics and Soldier–System Interface

Increasingly, it is difficult to consider the soldier in isolation from the systems he or she operates. There is an almost limitless set of potential applications of robotic systems that can be developed to equip the military of the future. These will range from systems in which the human is a controlling element—current unmanned aerial vehicles are an example—to fully autonomous systems that think and act independent of direct human control.

Not enough attention has been given to the role of the human in the loop in these systems, or the psychological consequences the human operator may experience as a result of their use. The Predator, an unmanned aerial vehicle, can be flown by a pilot who may be physically situated halfway around the globe from the vehicle itself. The psychological space that the pilot inhabits, however, may be in the area of operations. Moving into and out of the psychological space may be problematic. Psychologists need to study this more and develop strategies to aid UAV pilots in managing this aspect of their job.

Engineers are also closer to developing fully autonomous killer robots. These would be released on the battlefield and would kill enemy tanks, personnel, or aircraft without any direct human supervision. When I teach cognitive psychology to cadets, I ask them to extrapolate theories of human perception, attention, cognition, and decision making toward the design of the artificial intelligence system necessary to drive a fully independent robotic weapons system. Should we mimic human cognition, with all of its idiosyncrasies, or can human intelligence be improved upon in such a way to allow these systems to kill or otherwise execute their missions without error?

Psychologists will have their hands full in designing these autonomous robotic systems that mimic human thought and decision making, while at the same time training military personnel how to adapt to the very same systems. Technologies should amplify human abilities, not suppress them. When robots do the hard and most dangerous fighting, what happens to the role of the human? Will they become apathetic, fearful, and resigned to fate, or will they feel empowered by the capabilities of emerging systems? The psychological consequences of advancing robotic technology will have far reaching consequences on the profession of arms, morale, leadership, and the military ethic. Psychology will help solve the questions and will help ensure that more acceptable as well as combat effective systems will be developed.

Technologically Augmented Cognition

I am not sure if technology is making me smarter or dumber. Maybe you feel the same way. I am pretty good at math, but the computational skills I learned in public school, college, and graduate school have eroded. This decline began when I got my first hand calculator in 1975, and accelerated with the widespread adoption of personal computers. I remember doing complicated statistical procedures using a calculator in graduate school, some that took days to complete and to verify. Now I can do the same procedures in a matter of minutes using standard statistical software packages on my computer (which, of course, only takes seconds to do the actual computations). On the one hand, I can barely do fractions with a pencil and paper anymore. On the other hand, I can explore data sets with scores of variables and thousands of data points with the sheer click of a mouse.

And what about smart phones? My wife bought her first smart phone a few months ago. It is rarely out of her reach. I'll ask an idle question, like "What movie won the award for best picture in 1942?" and almost before the words leave my mouth, she will have the answer from her smart phone. Why would children—or any of us—memorize facts anymore, when any conceivable human knowledge is available from the smart phone? I remember memorizing the names of all 50 states and their capitals when I was in grade school. This seems like an unnecessary chore today. If you want to know the capital of Montana, just Google it!

These examples refer to an augmented memory system. Other aspects of human cognition can also be augmented, to include attention, perception, and decision making. Route finding, such as planning the best way to travel from your home to visit a friend in another city or state, used to require thoughtful effort. Now you just enter the address into your vehicle's GPS system, decide whether you want to listen to a male or female voice and perhaps what sort of accent it uses, and then drive off toward your destination. One less thing you have to occupy your mind with. But do you use the extra cognitive space to improve vigilance and care? Or does this simply allow you to text or talk on your phone instead of looking for road signs?

Extend these thoughts to the military setting. What do twenty-first-century soldiers or leaders need to know in their head, versus what is quickly obtainable through a device? Future weapons and command and control systems should be designed to let the human do what humans do best, and to let the automated system do what machines do best. The computer can process information faster than a soldier and has faster reaction times. But the soldier should be superior in higher order cognitive functions, like judgment.

Donald Norman, in his classic book *The Design of Everyday Things*, makes the distinction between "knowledge in your head" and "knowledge in the world."[17] The former refers to skills and facts that you must recall from memory, and the latter represents information that can be made easily available by the system you are operating. A simple example is fuel remaining in your car. A fuel gauge represents knowledge in the world—you can glance and see how much fuel you have left. This is much better than measuring how much gas you put into the tank, keeping track of miles driven, and then calculating how many more miles you can drive without refueling. The military engineering psychologist plays a critical role in helping designers decide what information should be in the head or in the world. Doing it right can have significant implications for system performance.

Properly designed future military systems will augment soldier performance by augmenting the soldier's cognition in the right way and at the right time. Even fully autonomous systems will be designed and programmed by humans. If we want those systems to perform like us, then we must understand our own cognitive architecture as we build it into the machines.

Neurobiologically Enhanced Cognition

The rapidly expanding field of neuroergonomics may allow scientists to alter brain functioning by direct manipulation of genetic, neuroanatomic, and neurochemical/ physiological systems. Leading neuroergonomist Dr. Raja Parasuraman describes four ways that neuroergonomics may allow scientists to augment different aspects of human behavior.[18] The first he calls *neuroadaptive interfaces*. We have already discussed this in the context of highly advanced prosthetics that may allow soldiers suffering amputations or other wounds that currently restrict their mobility and fluidity of movement to perform at levels equal to their preinjury levels.

Parasuraman describes three other applications of neuroergonomics that are relevant to the future military. *Molecular genetics*, based on the human genome, provides an ever-increasing level of understanding between genes, protein production, and the regulation of a host of psychological functions. Emerging methods that allow for the manipulation of gene expression may ultimately be used to enhance soldier attention, memory, and decision making, as well as allowing the soldier to respond more effectively to combat stress and adversity. *Multitasking and mental workload* also relates to neurological functioning, and may be improved through central nervous system interventions. Finally, Parasuraman identifies *human error* as a phenomenon that may be alterable through neuroergonomic interfaces. Fratricide, or the accidental killing of friendly forces on the battlefield, is a dramatic exemplar of human error in the military context. Reducing this sort of error through neuro-bioengineering could have profound effects on improving war fighter effectiveness and efficiency.

SUMMARY

Psychologists will serve as equal partners to engineers and physical scientists in designing ways to enhance soldier and system performance for the military of 2050. We can give soldiers the ability to fight for extended periods of time without sleep, to be resilient in the face of great adversity, and to kill more effectively when killing is called for. We can augment the physical and cognitive abilities of soldiers and create intelligent systems that, when meshed with the soldier or operator, exponentially increase combat power.

As an engineering psychologist, I sometimes remark that we exist to provide adult supervision to traditional engineers. I could extend this comment to the physical and biological sciences. What I mean by this is that new technologies and biological aids intended to enhance soldier effectiveness all have major psychological impacts. As all areas of military science develop, the role of the psychologist becomes increasingly central in making sure that emerging technologies truly augment soldiers. In the absence of psychologists teaming on an equal par with other experts, we run the risk of developing systems that at best are awkward to use and don't fully exploit their potential—and at worst, cause physical or psychological harm to those on whom these technologies are inflicted.

NOTES

1. Alice Calaprice, *The New Quotable Einstein*, 173.
2. Scales reviews Beyerchen's taxonomy and relates it to the overriding importance of psychology in contemporary war in Robert H. Scales, "Clausewitz and World War IV," S23–S35. The original article by Alan D. Beyerchen, "Clausewitz, Nonlinearity, and the Importance of Imagery," can be accessed at http://muse.jhu.edu/journals/international_security/summary/v017/17.3.beyerchen.html.
3. See Fritz Stern, "Einstein's Germany," 97–118.
4. See Frank J. McGuigan, *Experimental Psychology*, 2–4.
5. Estimates of the incidence of PTSD in combat vary a great deal and vary, no doubt, on many factors. One of the most informative sources I have found for understanding combat PTSD and resilience is Brian J. Lukey and Victoria Tepe, eds., *Biobehavioral Resilience to Stress*. The various chapters included in the book address the biological, behavioral, cognitive, and social factors that affect combat stress and resilience.
6. One indicator of the reception of CSF among army units is that in an early empirical test of the program's effectiveness, some units were given CSF interventions and other units, serving as a control condition, were not. Amazingly, the soldiers in one control condition unit wanted the training so badly that they hacked into the computer program running the interventions so that unit members could access the exercises. This ruined the experimental design, but this shows how strongly soldiers desire this sort of training.
7. For more information on this and related DARPA research, you may find the following article informative and provocative: Azeen Ghorayshi, "This Is Your Brain on the Department of Defense," *Mother Jones*, April 3, 2012, www.motherjones.com/blue-marble/2012/04/department-of-defense-neuroscience-bioethics-brains-law.
8. For a comprehensive technical review of the human stress response, see Florian Holsboer and Marcus Ising, "Stress Hormone Regulation," 81–109.
9. For example, see Richard A. Andersen, Eun Jung Hwang, and Grant H. Mulliken, "Cognitive Neural Prosthetics," 169–90.
10. These data are unpublished, but a briefing describing the study and its results was given to the Royal Norwegian Army Academy in April, 2009. You may contact me if you would like a copy.
11. Michael D. Matthews et al., "A Comparison of Expert Ratings and Self-Assessments of Situation Awareness During a Combat Fatigue Course," 125–36.
12. For example, see Greg Belenky et al, "The Effects of Sleep Deprivation on Performance During Continuous Combat Operations," 127–135.

13. Almost any introductory biopsychology book will provide the interested reader more information on the biology of sleep. I recommend James W. Kalat, *Biological Psychology*.

14. Dave Grossman, *On Killing*, 5–17.

15. Again, see Kalat, *Biological Psychology* for a thorough review of the genetic, biological, and neurochemical bases of aggression.

16. Trisha Meili, *I Am the Central Park Jogger*.

17. Donald A. Norman, *The Design of Everyday Things*.

18. Raja Parasuraman, "Neuroergonomics: Brain, Cognition, and Performance at Work," 181–86.

12

SPIN-OFFS: A BETTER WORLD
THROUGH MILITARY PSYCHOLOGY

When I started writing I wanted the best tools. I skipped right over chisels on rocks, stylus on wet clay plates, quills and fountain pens, even mechanical pencils, and went straight to one of the first popular spin-offs of the aerospace program: the ballpoint pen. They were developed for bomber navigators in the war because fountain pens would squirt all over your leather bomber jacket at altitude. (I have a cherished example of the next-generation ballpoint, a pressurized Space Pen cleverly designed to work in weightlessness, given to me by Spider Robinson. At least, I cherish it when I can find it. It is also cleverly designed to seek out the lowest point of your desk, roll off, then find the lowest point on the floor, under a heavy piece of furniture. That's because it is cylindrical and lacks a pocket clip to keep it from rolling. In space, I presume it would float out of your pocket and find a forgotten corner of your spacecraft to hide in. NASA spent $3 million developing it. Good job, guys. I'm sure it's around here somewhere.)
John Varley[1]

There is a story around that when the Russians learned that NASA had spent 3 million dollars developing the pen described above, they bragged that their writing instrument only cost pennies and worked flawlessly in the zero gravity environment. They used pencils.

A spin-off is a technology originally designed to support a specific purpose in a specific setting that later is discovered to have generalized benefit outside the domain for which it was originally intended. It turns out that the National Aeronautics and Space Administration takes much pride in spin-offs from the space program. They have a webpage devoted just to the spin-offs that accrue from the space program each year. Go visit it at http://spinoff.nasa.gov. You will find an impressive list of spin-offs from different phases of the space program that have been useful in general society.

The military also produces many spin-off technologies, but I don't think they do a very good job of marketing them. A good example is nuclear power. The same efforts that led to the building of the original atom bomb led to the development of nuclear power. Advances in nuclear medicine can be traced to the same source.

There are many other examples. In World War II, the military needed a rugged and durable vehicle that could be driven in harsh road and weather conditions. This resulted in the General Purpose Vehicle, later known better by its acronym, pronounced "Jeep." After the war, they became widely popular both in the United States and abroad, and served as the inspiration for several subsequent generations of sports utility vehicles. These vehicles remain, of course, very popular today. My 14-year-old niece wants one when she gets her driver's license.

Both aviation and the space program also grew exponentially because of the military. Although airplanes existed just prior to World War I, the exigencies of war drove the rapid development of airplanes from rudimentary platforms with very limited speed, maneuvering, and altitude capabilities to machines that could carry and employ heavy weapons, maneuver in dog fights, and fly at astonishing speeds (for the time). Modern rocketry was given a big boost by World War II, with the Germans firing V-1 rockets into England. These in turn were improved after the war, and ultimately evolved into rockets capable of placing satellites and humans into earth orbit and beyond.

Military spin-offs are not restricted to engineering and physical science. We have already seen that World War I led to the development of modern aptitude testing, for example. And World War II gave birth to engineering psychology. Clinical psychology also grew both in numbers, in response to the needs of millions of veterans, but also in scope, as psychologists learned how to better classify and treat the types of disorders frequently suffered by veterans.

I don't think the military does a very good job of letting the rest of us know how spin-offs from military technologies have benefited the world. I am sure that scientists and engineers of all types could fill volumes with the myriad of advances that stem directly from military funded research and development efforts. I hope that some senior decision maker in the Pentagon reads this chapter and is inspired to create something like NASA's spin-off page. The military needs to do a better job of promoting how the money it invests in science and technology provide return to its investors.

In the following pages I will speculate on how advances in contemporary military psychology may transfer to general society. These spin-offs will come from every area of military psychology and impact most substantive subareas of psychology. Many of these spin-offs will result from lessons the military is learning about the interactions of humans with engineered systems. Others will fall into the category of how psychologists treat psychopathologies. Still others will come from what the military learns about improving the adjustment and performance of pathology-free soldiers. Impactful spin-offs will also come from advances in leadership theory and practice and may help transform the workplace in future organizations.

IMPROVED UNDERSTANDING AND TREATMENTS
FOR STRESS-RELATED DISORDERS

Psychologists describe what are called potentially traumatic events, or PTEs. Over the course of a lifetime many people will experience a PTE.[2] These events are those that threaten a person's physical, emotional, or psychological well-being. They may be experienced directly or vicariously. Moreover, what is traumatic for one person might be something that another person will take in stride. A trained police officer may observe a mangled body and not be strongly affected, but the citizen who discovers the body may suffer lasting psychological trauma.

Years ago, when I was a reserve deputy sheriff, I was dispatched to investigate a deceased subject whose remains were found in a ditch alongside a rural rode. A mother and her young son had been walking home from the school bus stop when the boy saw something that he thought to be a dead pig lying in the ditch. He pointed it out to his mother, who saw it was a female whose body had been so mutilated that her son did not recognize it as human. The mother passed out. When I arrived I assessed the situation and began preserving the scene for the homicide investigators who would soon arrive to begin the formal investigation. Even though I was a seasoned law enforcement officer at the time and had seen many gruesome sights, this one was difficult for me to deal with. I can only imagine what impact the image of a severed head and mutilated body had on the unfortunate mother and her child. Certainly there is reason to believe that the impact on a young child could have enduring effects, since they lack the cognitive and emotional skills that most adults possess.

Over the course of a lifetime most people will experience a PTE. They may witness a violent death, such as resulting from a car accident, or be victims themselves of a violent assault. Natural disasters, extreme pain, a life threatening illness—all of these are PTEs. But whatever the specific event, they have in common a threat to the basic well-being of the person or their loved ones.

Most people end up coping effectively after exposure to a PTE. More intense PTEs result in longer lasting and more severe repercussions. Estimates vary but clinically significant pathologic reactions following PTEs are less than 10 percent. Of course more intense PTEs may have much stronger and longer lasting impacts on psychological adjustment. Witnessing the death of a stranger in an auto accident, while troubling for anyone, would likely be less disturbing than being involved in an accident where a loved one was killed.

Other factors are related to the impact of PTEs. Frequency of PTEs plays a role. Through sheer bad luck or by virtue of employment (e.g., law enforcement personnel, firefighters) some people may be exposed multiple times. Degree of training impacts the reaction to PTEs. Police officers, who train and prepare to respond to very dangerous and adverse situations, are often more resilient in dealing with PTEs than those who do not come into regular contact with PTEs as part of their job. People with preexisting pathology, drug dependence, or personal or social problems may be predisposed to more adverse reactions to PTEs than better-adjusted individuals.

The military represents a somewhat unique milieu in which to study human reactions to PTEs. Unlike most civilians who may experience one or two PTEs through their lifespan, combat soldiers may experience many PTEs through the course of a deployment. Seeing fellow soldiers die, viewing dead and wounded civilians, and personally killing another human being represent situations that soldiers experience regularly. Besides the obvious PTEs like those just described, simply being displaced for months at a time from their family and life's normal routines, coupled with the ever-present possibility of an enemy attack, is itself a PTE. And there are existential aspects of combat that may be construed as PTEs. Observing orphaned children, witnessing graft and corruption, and questioning the meaning and purpose of a given mission or deployment can be traumatic. In short, soldiers experience a world that is incomprehensible to most civilians.

Depending on the situation and how pathology is defined, combat may result in PTSD rates of 15 percent or higher.[3] This means that in the current war tens of thousands of soldiers have experienced a pathologic response to PTEs and require treatment for their psychological injuries. Unlike civilians, who may be exposed to a PTE and not report it to their physician or other health care professional, the military has in place mechanisms to help soldiers who experience significant psychological trauma. It is from them that new methods of treating stress related disorders comes. If the Vietnam War was the stimulus to include PTSD as a diagnosable psychological disorder, the wars in Iraq and Afghanistan have fueled the development of methods for more effectively treating this and related combat stress disorders.

The military is also involved in basic behavioral neuroscience aimed at identifying the biological basis of depression. With historically high levels of suicide and depression among soldiers, it is critical to isolate and study possible biological markers. Our laboratory at West Point is currently engaged in a fascinating collaboration with Nobel Prize laureate Dr. Paul Greengard and his associates. Scientists in Greengard's lab have identified a protein, dubbed p11, that is linked to depressive behaviors in mammals. Mice depleted of p11 show depressive behaviors and high p11 is associated with normal adjustment. Common antidepressant drugs—especially those that target the neurotransmitter serotonin—can compensate for lower p11 levels. Also, and this is very interesting, common over-the-counter nonsteroidal anti-inflammatory drugs (NSAIDs) counteract the effects of p11 on serotonin levels.

At West Point we are testing p11 levels in a sample of cadets with the aim of linking these levels to various measures of adjustment. Given the stressful nature of life for

West Point cadets, this represents a chance to see how p11 might be linked to meaningful behavioral and emotional outcomes among military personnel.

There is another intriguing aspect of this research. Military personnel are heavy
users of NSAIDs. Soldiers are essentially professional athletes who never have the luxury of an off-season to heal their aches and pains. Accordingly, they self-medicate with
NSAIDs and other over-the-counter medications in great numbers. Remember, there
is good evidence that NSAIDs counteract the beneficial effects of antidepressant drugs
that aim to maintain normal serotonin levels in the brain. Soldiers are often prescribed
antidepressant drugs that maintain serotonin levels. But if they take NSAIDs at the
same time, then they are undoing the beneficial consequences of the antidepressant.
Wouldn't it be amazing if we discovered that the increased suicide rates among military
members are related to their use of pain medications available without a physician's
prescription? These NSAIDs include instructions and information about drug interactions, but most people do not read them.

Desperate times call for desperate measures, and the military is quick to experiment with new ways of treating PTSD. I expect new and improved drug therapies for
PTSD, first tested on soldiers, will find their way to the civilian sector. Depending on
our West Point study and others like it, novel treatments that aim to enhance p11 in
soldiers or simply advise those with depressive illness not to take NSAIDs will emerge.
Other treatment approaches that are being tried include hyperbaric medicine, mindful
meditation, and yoga. These emerging treatment protocols are receiving a lot of military funded research support. There is evidence that mindful meditation may result in
increased concentrations of brain derived neurotropic factor (BDNF), a neurochemical that promotes increased density and complexity in the connections among brain
cells.[4] Perhaps drugs that promote BDNF may be derived that help soldiers recover
from combat-related stress disorders. Time will tell.

A less obvious spin-off is also occurring. The military is fighting hard to destig-
matize mental health problems. Through massive education efforts like the Army's
Comprehensive Soldier Fitness program, soldiers are becoming more willing to seek
help for psychological issues. Once considered a sign of personal weakness, soldiers
are coming to understand that acknowledging their psychological wounds and having
them treated is better for them and their units than suffering in silence.

The reluctance to seek help for psychological problems is unfortunately not
restricted to the military. It is also true in civilian society, especially among men. The
personal and economic costs to society are profound. People who refuse to seek psychological treatment suffer needlessly, may turn to alcohol and drugs to self-medicate,
or even commit suicide. Thus, the lessons learned from the military in how to des-
tigmatize mental illness may be adopted as part of a national health care strategic
policy. Such efforts, if successful, would have direct positive impacts on critically

important outcomes ranging from reduced substance abuse to increases in workforce productivity. And along the way millions of people would live more enjoyable and meaningful lives.

PROMOTING RESILIENCE

Remedying disease is not the same thing as promoting resilience, and promoting resilience may stand to be the most significant spin-off from twenty-first-century warfare. In the United States, the term "health care" is actually a misnomer. Really, we have a "sickcare" system. A sickcare system responds only to diseases and pathology, and offers little in the way of preventing disease and promoting healthy life styles. If you are lucky enough to have health insurance and access to decent medical facilities, then our system does a competent job of treating diseases and injuries. But many millions of Americans still do not have access to insurance or adequate medical care.

Moreover, the absence of pathology is not the same thing as health. Historically high rates of obesity, lack of exercise, poor nutrition, use of both licit and illicit drugs, and bad sleep habits compromise the wellness of tens of millions of Americans. Left unaddressed, these conditions and behaviors lead to heart disease, cancer, high blood pressure, diabetes, and a host of psychological problems including depression, anxiety, and even suicide.

It is equally true in psychology that absence of pathology does not equate with wellness and optimal functioning, as the positive psychologists have begun to tell us. People cope with problems that may not meet clinical criteria for a specific diagnosis but impair the quality of their personal and social lives. But psychology itself has long followed the disease model. Clinical psychologists get paid to identify, diagnose, and treat disorders. They do not get paid to improve functioning among those who would truly benefit from behavioral and psychological change that would improve their life quality.

This disease model mindset permeates the medical and psychological care systems. Imagine a health continuum (broadly defined to include physical and mental health) ranging on the far left of disease/disorder to the far right of physical fitness and psychological flourishing. The current system encompasses only the left end of the spectrum. What would be the effects of broadening the focus of our medical and psychological care communities to include the entire continuum? People would not only live longer lives, they would also live more productive and fulfilling ones. Improving the physical, emotional, psychological, and social adjustment of millions of people—in addition to continuing to provide medical and psychological care where needed—would ultimately extend the life span, reduce drug and alcohol abuse, result in a reduction of crime, and improve national productivity.

I am not alone in coming to this conclusion. Dr. Richard Carmona, the former Surgeon General of the United States, was present in the meetings with the Army Chief of Staff when Dr. Martin Seligman and I explained the rationale of positive psychology and how it could be used as the conceptual basis for developing a true health care and psychological fitness training system for the army. Dr. Carmona was vocal about the need for a parallel approach in the health care system for the nation as a whole. A national health care policy based on a holistic approach to both disease and wellness, he argued, would result in untold benefits to the nation of the sort just mentioned.

The Army's Comprehensive Soldier Fitness program is revolutionary in its approach to psychological well-being.[5] It is revolutionary in its focus on improving emotional, social, family, and spiritual fitness among soldiers, and also in the technologies and strategies it is developing to achieve this aim.

A key spin-off from CSF may be the comprehensive resilience modules, or CRMs, that are being developed to enhance personal resilience. When a soldier completes an online fitness evaluation, he or she is given feedback on their emotional, social, family, and spiritual fitness. If they are low in any of these areas, they are directed to a menu of online fitness development training exercises. These exercises are evidence-based; that is, they have been shown to improve fitness and resilience in empirical studies. For example, a soldier who scores low in emotional fitness may choose to take a CRM based on Seligman's "three blessings" exercise. In this exercise, the soldier is asked to record, at the end of each day for a week, three things that went well that day and why they went well. Evidence shows that people who complete this simple exercise are happier and show fewer depressive symptoms than people in control groups who do not complete the exercise.[6]

This is just one of numerous CRMs developed under the CSF program to improve overall psychological adjustment, instead of treating a psychological disorder. Over time, the army will determine which CRMs are effective and which are not. New ones will be added and validated. Other dimensions of resilience and fitness may be identified and training modules will be designed specifically for them.

Spinning off CRMs to the general population would represent a historic contribution of military psychology to the nation as a whole. I think people will embrace these methods. Note the widespread interest in computer-based mental exercises designed to improve cognitive functioning. If people see the need to improve attention and memory, they are also likely to see the value in improving their own resilience. It is a tough world out there, and many people will jump at the chance to build skills to make them both better adjusted and more successful.

CRMs are not the only product stemming from CSF. The Global Assessment Tool (GAT) that is used to measure psychological fitness in soldiers can be directly exported to the general society. By taking this simple test online, people can self-assess areas where they are currently strong and robust and areas where they could improve their psychological fitness.

Other institutions can leverage resilience training procedures being used or under development by the military and apply them to their own populations. I can imagine college counseling centers adopting CSF technologies to help students adjust to college life. Large companies may employ them to help their workers. High-stress organizations like police departments and hospitals will benefit tremendously from adapting these methods. The possibilities are endless.

SELECTION

Remember that large-scale psychological testing began when the United States needed better methods for selecting people to serve in the military during World War I. The wars of the twenty-first century may fuel a revolution in selection technology.

The spin-off of improved personnel selection will be of obvious benefit outside of the military. The greatest improvements will come from better ways of testing noncognitive factors that are linked to employee performance. Personality, character strengths, emotional fitness are all emerging as factors critical to soldier performance. The positive psychology movement has driven vast improvements in assessing these contributing factors to job performance. Tests like the Army's Global Assessment Tool, which measures emotional, social, family, and spiritual resilience, may be easily modified for use in corporations. University of Pennsylvania psychologist Angela Duckworth offers cogent arguments that aptitude, in the form of intelligence and other traditional psychological testing, accounts for only a fourth of human performance.[7]

Other selection methods may also emerge. Advances in genetics that may aid in the selection and assignment of soldiers will transfer to nonmilitary occupations. Immersive simulators used to train soldiers may also be used as selection devices to assess how potential recruits naturally react to highly stressful situations. Some people are not cut out to be warriors, but may still be useful and productive soldiers if assigned to noncombat jobs. Similarly, advanced simulations may help industry select individuals who possess the right stuff for the types of jobs they might occupy.

The outcome is that selection methods initially developed by psychologists for the military—be they psychological tests, genetic or biological markers, or immersive simulations—will spin off outside the military and enhance the capability of other institutions to more effectively select their members. This selection revolution, by matching people to jobs for which they are optimally suited, may have tremendous positive impacts on national productivity and individual job satisfaction for millions of workers.

TRAINING

Organizations struggle to train employees in critical job skills. This may be especially true in high-stakes professions such as law enforcement and firefighting. And the simple how-to skills are often not the ones that organizations have trouble teaching. Law enforcement academies do a good job of teaching new officers how to fire weapons, restrain suspects, and drive vehicles under emergency conditions. More challenging is the goal of teaching higher order cognitive skills such as decision making and judgment.

The military is leading the way in using virtual simulations to train its personnel. As we discussed earlier, military aviation simulation is very advanced. It trains new pilots to high skill levels and helps experienced pilots sustain and improve those skills. Advanced aviation simulators are now integral parts of civilian aviation training. These simulation technologies are now being expanded for use in other military milieus including infantry operations.

Developing these simulation capabilities is not simply a matter of more realistic representations of the battle space and key tasks. Making trees and buildings look like the real thing helps, but the most important advances are in knowing *what* to train and *how* to train it. The "what" question involves determining the types of tasks and skills that simulators are most effective in teaching. In training soldiers on rifle marksmanship, there may be no substitute for traditional training with real rifles and ammunition at a range. But once the soldier has mastered the mechanics of rifle marksmanship, virtual simulations may be more effective in teaching the soldier when and under what conditions to fire the weapon. In short, simulations may be best utilized to teach the higher order cognitive skills mentioned above.

The "how" part of simulations refers to the manner in which training is presented in the simulation. The order, complexity, and speed of simulations developed to train small unit leadership skills and decision making must be well understood in order to maximize the effects of training. There are some great computer games on the market that involve firing various weapons against enemy soldiers. They are quite entertaining, but without a larger training plan a soldier could play the game for hundreds of hours and not learn anything useful. So how material is presented, how feedback is given, how learning is assessed, and how it is generalized to real operations is critical.

The goal is to enhance traditional training to the point that new soldiers and leaders can enter into combat operations as close to maximal performance levels as possible. Military aviation has achieved this goal, and over the next few years similar success will be found for infantry and other military occupational specialties.

The spin-off here, then, is to other professions where effective decision making, judgment, and other higher order skills may take years of on-the-job experience to develop. I can speak from my own personal experience that no matter how good traditional classroom-based law enforcement training may be, the first day you drive out on patrol by yourself you will immediately realize that it was not good enough. If every new law enforcement officer could run through one thousand well crafted virtual simulations before hitting the streets, both the best interests of the officer and the public he or she serves would be served.

Most law enforcement commanders—and the civic leaders they report to—will tell you that they worry most about poor judgment and decision making in their officers. Bad decisions that occur during vehicle pursuits or the employment of deadly force result in significant financial and legal woes to civic agencies.

Currently—where available at all—law enforcement simulations mostly focus on deadly force and vehicle operation. The so-called shoot–don't shoot simulators provide great preparation for real situations where an officer must decide, often in just seconds, whether or not to use deadly force on a suspect. Expanding this virtual training to hostage negotiations, handling domestic disputes, dealing with suicidal subjects, negotiating with people from diverse cultural backgrounds, and a myriad of other situations that law enforcement officers find themselves in would improve job performance and lessen liability for the community.

It is easy to see how these simulation technologies could also extend to fire departments, first responders, and other agencies that require fast and accurate decision making, sometimes under conditions where lives are on the line. These same technologies may also be invaluable in other venues as well. High stakes does not always equate to loss of life or bodily injury. Corporations where bad decisions cost jobs or huge sums of money are also venues where cognitive trainers, based on those developed by and for the military, may help managers make faster and more accurate decisions that will benefit the corporation.

This spin-off may ultimately result in a revolution in public and private sector training. This will impact national productivity, the effectiveness of public sector services, and the personal welfare both of the people serving in the organizations and their customers.

COGNITIVE DOMINANCE

Most jobs require us to be mentally quick and agile. We need to know how to allocate attention to our most pressing task, have fast and accurate access to memories, and quickly evaluate situations and render fast and accurate decisions. Moreover, we want to preserve these mental abilities as we age. With extended life spans, our enjoyment of later years will be directly tied to our cognitive abilities. Cognitive dominance refers to the ability of soldiers to harness component cognitive skills in order to make rapid and accurate decisions under high-threat conditions.[8] These attributes are critical to the military. Military leaders are becoming interested in how to improve and sustain high-level cognitive skills among its members. Cognitive dominance and fitness is coming to be seen as of equal importance to physical fitness in the military.

How can one train, improve, and sustain cognitive skills such as attention, perception, working and long-term memory, and decision making? How can metacognitive skills—that is, the ability of the person to utilize cognitive subskills to accomplish tasks—be trained and developed?

Interest in these questions is not confined to the military. There are commercially obtainable computer programs available now that promise to increase cognitive skills through practice. Advertisements proclaim that engaging in these mental exercises will improve various aspects of cognition. Such claims may pass the common sense test. If push-ups make your arms stronger, why wouldn't practicing memory drills improve the size and speed of working memory, for instance?

The psychological science behind this is still in its infancy. It is not yet certain that mental gymnastics truly improve cognition. But as you may suspect, the military is keenly interested in this topic. High stress, restricted sleep, and the adversity of being away from friends and family can impact cognition. One approach is to develop training protocols that soldiers may use to remain mentally effective. Toward that end, the military is supporting systematic research on how to improve and sustain these skills through practice. The Defense Advanced Research Projects Agency (DARPA) is actively funding research to support this goal.[9]

Although this research is just beginning, the military, with its vast financial resources and access to leading scientists, stands in a favorable position to determine whether cognitive muscles can be strengthened like physical muscles—and, if so, how to best achieve this goal. I would classify this as a potential spin-off from military psychology. If valid and reliable cognitive training protocols can be developed for soldiers, they will have massive applicability to general society. President Kennedy was the champion of physical fitness in children. Imagine a cognitive gym class for grade school children, where they spend 30 minutes a day strengthening their memory or other cognitive skills! At the other end of the spectrum, senior citizens could use these methods to sustain long into their life span their ability to think, remember, and decide. In between, workers of all sorts could improve productivity by strengthening these skills.

PROSTHETICS AND BRAIN-MACHINE INTERFACE

As we saw in the previous chapter, tremendous gains are being made in prosthetics. Based largely on advances driven by the flood of amputees into military hospitals, vastly improved artificial limbs—wired directly to the patient's own brain—will revolutionize the future prospects of amputees, both military and civilian. These limbs, which behave and feel like the real thing, will allow untold numbers of people retain nearly the full function and associated range of activities associated with their own limb.

The emerging field of brain-machine interface (BMI) will enable the development of a myriad of applications besides prosthetics. Workers may be able to control complicated machines and systems directly through brain activation. Pilots may control selected subsystems of their aircraft through BMIs. Paralyzed people will use BMI to control their world, perhaps even freeing them to operate vehicles or other systems they may, depending on the nature of their paralysis, currently be unable to control.

LEADERSHIP

There are few human enterprises where leadership matters more than in the military. In preparing for and fighting war, effective leadership may mean the difference between victory and defeat, and between life and death. No institution pays more attention to leader selection, development, and behavior than the military. Unlike most other institutions, the military relies on systematic research, funded through its various labs and research organizations, into the scientific basis of leadership.

In contrast, private sector and public agencies and institutions pay a lot of lip service to leadership. To paraphrase the old comment about the weather, everyone is talking about leadership, but nobody is doing anything about it.[10] Bookstores are well stocked with best selling books by so-called leadership experts, but for the most part the authors' expertise is much more in the area of creative writing than in systematic and empirically validated leadership theory and practice.

I suspect that your experiences with leadership in the organizations you work for reveal a need for improvement. Part of the problem may be that nonmilitary organizations hire managers, and that is just what they do. They manage budgets, schedules, rules and regulations, use of vacation and sick time and so forth, but they do very little leadership. You probably should not blame your manager for managing. She is, after all, just doing what her boss expects. And for managers, like most workers, this is just a job that starts when they arrive at work and ends when they depart at the end of the day.

To the military leader, the task of leading others is a 24/7/365 enterprise. It is more than a job—it is their profession. Leadership involves many tasks in common with management, but it goes much farther. It requires the leader to develop cohesive, high-performing teams. The military leader must inspire others through his or her own personal behavior and character. They must respect and trust their subordinates, and always place the well-being of their soldiers above their own personal needs. By supporting and empowering those around them, the entire organization benefits.

So what will be the leadership spin-offs from twenty-first-century war? First and foremost, the lessons being learned in the emerging field of in extremis leadership should be very relevant to nonmilitary organizations. Some of these organizations, like law enforcement agencies and fire departments, bear significant similarities to the military in terms of the sorts of occupational stress and personal risks that workers encounter. I believe that in extremis leadership principles also generalize to corporate settings. Lives may not literally be on the line, but livelihoods certainly are. Bad leadership can cause irreparable harm to corporations, impacting profits and putting jobs and even the survival of the company at risk.

Those who work in poorly led and poor performing corporations face many of the same psychological risks that combat soldiers encounter. They are not at enhanced risk of PTSD, but in these low-functioning organizations the rates of other significant psychological problems abound. Depression, substance abuse, suicide, and anxiety can be direct consequences of poor leadership. Many cases of workplace violence, from simple assaults to the murder of coworkers can stem from poor leadership.

Exporting Tom Kolditz's principles of in extremis leadership to the corporate world may help these organizations select and develop true leaders, not just managers.[11] Fortune 500 companies already spend millions of dollars in leader development, but much of it is not well grounded in behavioral science.

Of course, not all leadership occurs in dangerous contexts. *Mundane* leadership, that which occurs in organizations where lives, livelihoods, or reputations are not so regularly on the line, is also important. Military leadership experts argue that leadership—not just management—is also critical in such organizations. Good leadership may set the occasion for workers to enjoy greater personal growth, high job satisfaction, and increased self-esteem. Just because lives are not at risk is no reason to subject these workers to management instead of leadership. The principles of leadership as laid out in the army's leadership doctrine, Field Manual 6-22, can easily be adapted to mundane leadership, just as emerging principles of in extremis leadership may be extrapolated to higher stakes, nonmilitary organizations.[12]

MULTICULTURAL SKILLS

Every major social institution including business, education, and, of course, the military knows that its members must be culturally literate to perform effectively in the twenty-first century. This is not a new revelation, to be sure. Technological advances of the twentieth century such as fast, cheap, and convenient air travel between continents began to shrink the world and break down traditional barriers between countries and cultures. The computer and the Internet further compressed the world. And the universal availability of smart phones may have provided the coup de grace to the segregated and fragmented world of the nineteenth and early twentieth centuries.

Social media allows people to communicate on a personal basis, in real time, from virtually anywhere in the world. Many of us have Facebook friends across the globe. Teens and adults alike spend many hours a week tweeting and posting. In an activity called "catfishing," some young people develop romantic relationships online with a person they have never met in person.[13] Some parents may wonder whether their children have living, breathing true-life friends anymore. In our fluid society, we can easily stay in instant touch with our friends, associates, and former colleagues following moves to a new city, state, or country.

The implications are clear. In a tightly interwoven world culture people must learn to appreciate and be sensitive to cultural values, practices, and beliefs that are different from their own. This is not a new fact, but its impact is more universal now. I am sure that generals, heads of state, and business leaders have always needed to possess these skills. If they didn't they would lose wars, alienate potential allies, or fracture business alliances. But now every member of the institution has the potential for frequent and meaningful contact with people from vastly different cultures. Right or wrong, judgments about the many are often made from contact with the few.

We have seen how this plays out in the military in previous chapters. A young infantry private who violates an important cultural norm while on deployment may bring severe consequences on his unit. Widespread apathy or blatant disregard of local cultural traditions may ultimately turn the local population against the invading military (or an occupation force) with resultant failure to meet strategic objectives.

By extension, the same is true in other institutions. Universities may recruit heavily for students from other countries, but if these students find the social and cultural environment at their university to be unacceptable they will simply leave, and other potential recruits may avoid that particular institution. In an era where many colleges struggle to fill out an entering class, this represents loss of revenues and a threat to the continued existence of the institution itself. And even if the institution survives, the potential loss of cultural diversity will degrade the overall quality of the educational experience. Business executives who work hard to establish good relationships and diplomacy with customers from diverse cultures may see their efforts undermined by lower echelon employees—the ones who do the real work within the corporation—if those employees are not also well skilled in cultural matters.

Better and more effective multicultural skill education and training will, therefore, be an important spin-off from current and future wars of the twenty-first century. Through hard experience, the military has learned that cultural competency is more involved than learning a second language or simply spending time in a different culture. Contemporary efforts developed by the military to train these skills include the latest advances in simulation technology. Enter relevant keywords into your search engine, and you will discover the great scope of research sponsored by all branches of the US military on games and simulations designed to improve cultural skills.

I have focused many of my comments on how soldiers interact with local populations or the enemy. It is equally important to see what the military is doing to train its leaders to work effectively with their counterparts from other countries in multinational operations. The games and simulations emerging from these research and development projects should have tremendous value in helping strategic and operational level leaders within the corporate world learn to interact more effectively with their counterparts around the world.

This spin-off may ultimately be among the most important ones that emerge from military psychology in this century. The need and motivation to improve cultural understanding is clearly present and is simply waiting on the development of truly effective and efficient methods for conducting this training. The military, where cultural understanding is literally a life-or-death matter, is leading the way in innovative training in this domain. The benefits to business in particular stand to bolster the world economy and improve cooperation and productivity for all countries involved.

CONCLUDING THOUGHTS

Benjamin Franklin wrote, "I am apt to think there has never been, nor ever will be, any such thing as a good war, or a bad peace."[14] I am sure you join me in agreeing with this sentiment. Still, it is in our human nature that adversity drives adaptation and innovation. All aspects of human invention—science, medicine, literature, and the arts—are inextricably and irrevocably linked to war. Detesting war and wishing to avoid it should not blind us to the spin-offs from war that improve the human condition.

Advances in military science are not limited to improved deer rifles or better and more rugged off-road vehicles. In almost every domain war has driven scientific developments that have improved the lives of all people. Modern approaches to treating malaria, for instance, grew from the efforts of military physicians sent to address an epidemic of malaria impacting workers digging the Panama Canal.[15] Are you a backpacker who appreciates easily portable prepackaged foods? You can, at least partially, thank the military who developed C-rations, K-rations, and, more recently, meals-ready-to-eat (MREs). Camouflaged clothing—essential to turkey hunters–owes much of its development to military researchers. Former West Point engineering psychologist Dr. Timothy O'Neill began using his knowledge of the human visual system to design modern digital camouflage over 30 years ago. Today, he consults with various clothing designers to make this state-of-the-art camouflage available to hunters, law enforcement personnel, and others with a need to conceal themselves from easy visual detection.

The contributions of military psychologists are often not as obvious or sexy as those stemming from other sciences. But you can bet that millions of people will benefit directly from the spin-offs that come about directly from psychological research and practice that addresses the most pressing needs of the military. From military psychology we will improve the quality of the personal and vocational well-being of virtually every human being, be it through better prosthetics for amputees, resilience skills for crime victims, or more effective leadership in our corporations.

NOTES

1. John Varley, *The John Varley Reader*, 271.
2. The incidence rate of PTE exposure varies somewhat across nations, but in the United States appears to be a lifetime incidence rate of 75 percent or so. See Ronald C. Kessler et al., "Posttraumatic Stress Disorder in the National Comorbidity Survey," 1048–60.
3. Estimates of combat-induced PTSD rates vary a good deal, depending on the type of combat experience, degree of PTE exposure, and type of unit the soldier is assigned to. The following article reports incident rates for marine corps members, but the introduction describes incidence rates in other services and settings: Gerald E. Larson, Robyn M. Highfill-McRoy, and Stephanie Booth-Kewley, "Psychiatric Diagnoses in Historic and Contemporary Military Cohorts," 1269–76.
4. Glen L. Xiong and P. Murali Doraiswamy, "Does Meditation Enhance Cognition and Brain Plasticity?" 63–69.
5. Martin E. P. Seligman and Michael D. Matthews (eds.), *American Psychologist Special Issue*.
6. Martin E. P. Seligman et al., "Positive Psychology Progress," 410–21.
7. Angela L. Duckworth et al., "Grit," 1087–101.
8. I coined the term cognitive dominance several years ago to refer to the ability of military leaders to make decisively better and faster decisions on the battlefield than their opponents. Working with retired Army Major General Bob Scales, we worked toward the goal of developing immersive cognitive skills training that would quickly give small unit combat leaders the edge in outthinking the enemy.
9. Ralph E. Chatham, *A Possible Future for Military Training*.
10. The weather quote ("everybody talks about the weather, but nobody does anything about it") has been attributed to Will Rogers, Mark Twain, Charles Dudley Warner, and other sources. A convincing argument in favor of Charles Dudley Warner is made at the Quote Investigator website; see http://quoteinvestigator.com/2010/04/23/everybody-talks-about-the-weather/. Additional discussion of the quote can be found in Paul F. Boller, Jr. and John H. George, *They Never Said It*, 124.
11. Thomas A. Kolditz, *In Extremis Leadership*.
12. US Department of the Army, *Army Leadership*.

13. A notable case of catfishing involved Notre Dame football player Manti Te'o, who established a relationship with a woman and suffered through her death, without having ever met her. This may seem wildly incredible to someone of an older generation, but my cadets tell me that online romances are common among their peer group.

14. Quoted in H. W. Brands, *The First American*, 620.

15. Christian F. Ockenhouse et al., "History of US Military Contributions to the Study of Malaria," 12–16.

13

PSYCHOLOGY, WAR, PEACE, AND ETHICS

I think that people want peace so much that one of these days government had better get out of their way and let them have it.
Dwight D. Eisenhower[1]

Psychology is the science of human behavior. Even psychologists who exclusively study the behavior and physiology of animals have, as their ultimate goal, the desire to improve our understanding of human nature. As social animals, humans form groups that convey survival advantage to their members. Alone, we are not a particularly imposing animal. We are slow and clumsy and lack large teeth, horns, poison glands, and the like that other animals may employ to obtain food or protect themselves. But we are an intelligent species, and, when formed into groups, we can accomplish things no other animal can come close to achieving. Over time small informal groups of humans who subsisted on hunting and gathering developed into larger, more sophisticated collections of humans. In the end political entities emerged in the form of cities and nations.

Inevitably, groups compete for resources. Primordial bands fought over the best hunting grounds or access to the best fishing. As nations emerged, they came to clash not only over natural resources, but also over ideological and political issues. As these groups became more formally organized and sophisticated and as technology continued to develop, the phenomenon of war appeared. The concept of war formalizes and legitimizes the use of violent aggression between nations.

War, then, is a social behavior supported by the individual behavior of members of the warring groups. By definition war is essentially a matter of human behavior, and therefore a psychological phenomenon. Whether combatants attack each other with sticks and stones or high-speed aircraft, the essence of war boils down to a question of human behavior.

By the same reasoning peace also reduces to psychology. Peaceful behavior is more than the absence of war. It includes individuals and groups working together for the common good, social cooperation, empathy, and consistent prosocial behavior. In the same way that military psychologists must focus on resilience and human flourishing in addition to disease and pathology, they must also apply their knowledge of the science of human behavior to encourage peaceful behavior, not just an understanding of war.

In this concluding chapter I briefly review the role of military psychologists in supporting war. Then, in more depth, I discuss what the discipline of psychology offers in promoting peace. I conclude with thoughts about the ethical challenges military psychologists face in their support of the military.

PSYCHOLOGY AND WAR

During my first college course in psychology, I decided that I wanted to become a learning theorist. B. F. Skinner's book *Beyond Freedom and Dignity* had just appeared.[2] Skinner challenged psychologists to focus on the science of behavior and to extend that science to the most important spheres of human activity. The writings of Clark Hull, Kenneth Spence, E. G. Guthrie, and Edward Tolman also captured my imagination. In due course I based both my masters and doctoral dissertations on fundamental aspects of conditioning and learning.

Little did I imagine that I would soon morph into a military psychologist. Nearing the completion of my doctorate of philosophy from the State University of New York at Binghamton, I joined the air force and transitioned from studying the behavior of rats in mazes to the selection and training of military personnel. This was a revelation. The basic science I had learned in graduate school had much to offer in applied settings. To my surprise, conducting research of this sort was intellectually challenging and had the bonus of having the possibility of helping improve the lives of other people. A military psychologist was born.

My story is similar to that of many military psychologists. To my knowledge no universities offer doctoral degrees in military psychology. Most psychology graduate students don't start out intending to become military psychologists. But what we learn in completing our graduate degrees is that no matter what specific subarea of psychology we are in, it has a home for application in the military context. Clinical psychologists easily switch their attention from civilian populations to treating soldiers. There are no unique set of disorders found only among soldiers; hence there is no need for graduate programs in military clinical psychology. Similar things can be said about all other psychological specialties in which people obtain degrees. Human behavior and human nature simply occur in different contexts, so education and preparation in our chosen subdiscipline of psychology, whatever it may be, is sufficient.

As we have seen, psychologists have made significant contributions to the military since the establishment of psychology as an independent discipline. And as the nature of warfare and its supporting technology have evolved, the role of psychology as a difference maker in war has been increasingly recognized by the militaries of nations around the world. I do not know of any modern military that does not have formal and systematic ties to psychology. As I have traveled around the world, consulting and collaborating with military psychologists from a culturally and politically diverse array of nations, I am astounded by the observation that we are all working on solving the same problems.

Division 19 of the American Psychological Association (APA), the Society for Military Psychology, was one of the charter divisions of the organization. Psychology is more relevant and viable today for the military than at any point in history. Hundreds of psychologists work directly for the military either as uniformed officers or federal civilians. And many more psychologists support the military through contract work for the US Army, Air Force, Navy, or Marine Corps. The *Oxford Handbook of Military Psychology* offers a comprehensive review of contemporary military psychology. A glance at its table of contents will reveal the breadth of relevance of psychology to today's military.[3]

It is clear that psychology will play an even more important role in the future for the military. Social, political, and technological changes necessitate that the military capitalize on the knowledge possessed by psychologists to ensure that the military of the future will be combat ready, resilient, and lethal when needed. To get the most out of its soldiers and systems, the military must aggressively incorporate state-of-the-art psychology into all aspects of its missions.

The career of a military psychologist can be quite an adventure. I had anticipated a traditional academic career: obtaining tenure at a traditional university and spending an idyllic life—probably rooted down in one place—teaching, conducting research, and consulting. But my unexpected plunge into military psychology allowed me to tackle an ever changing series of fascinating problems with real life implications; live and work at four different military installations in four states; visit and collaborate with other military psychologists across North America, Europe, and Asia; and even teach at the college level within my original academic areas of expertise—learning and biopsychology. My military psychologist colleagues have done this and much more, with many deploying to war zones. Compared to the sometimes stuffy and predictable world of traditional academia, this has been an almost Indiana Jones sort of ride. If you are a young psychologist or psychology student looking for an exciting alternative to the more traditional career paths, I recommend getting involved in military psychology. It is a growth industry, and one that provides large measures of personal and professional satisfaction. You can help bring to fruition some of the futuristic ideas I have presented in this book.

PSYCHOLOGY AND PEACE

The very same basic psychological science that can aid the performance of the military in its missions can also be used to promote more peaceful behavior. Division 48 of the APA represents a diverse collection of psychologists who share the vision of applying these principles to promote peace, decrease conflict, and end violence. Called the Society for the Study of Peace, Conflict, and Violence, Division 48 provides a reservoir of talent and motivation to further the attainment of these objectives. Its journal, *Peace and Conflict: The Journal of Peace Psychology,* provides a forum for presenting scientific findings and theoretical perspectives on the psychology of peace.

After the United States deployed its military forces in Afghanistan late in 2001 and in Iraq in March 2003, the importance of both military and peace psychology was greatly enhanced. After 2004, when it became apparent that these would be long wars with little chance of a traditional political or military victory, Divisions 19 and 48 began to cooperate more than ever before to address common areas of concern.

At that time I was just becoming active in Division 19 leadership. During the annual APA convention, we would host a hospitality suite that included opportunities for informal dialogue among our members (and, yes, ample chances for the consumption of potent potables). One year, a representative of Division 48 asked if they could send a delegation to visit the Division 19 leadership in the hospitality suite. We thought it was a terrific idea. But to be frank, we weren't sure quite what to expect. By this time, the atrocities of Abu Ghraib had become widely publicized and had elicited a good deal of antipathy toward the military and, at times, toward military psychologists.

A small group of Division 48 psychologists visited our hospitality suite as planned. At first both groups seemed uncomfortable. We didn't want to be labeled as warmongers, and I imagine they didn't want to be labeled as naive peaceniks. I think we looked at each other like we were meeting aliens from another planet. But I could not have been more satisfied with the outcome of the meeting. Within a short while we were quickly discovering that as psychologists, we shared a vision of a peaceful world. They were glad to hear that military psychologists do not support torture, and we were relieved to learn that the peace psychologists understood the unfortunate political realities of the modern world that lead to war. We stood on common ground, with similar views, and got along great personally. Maybe the wine helped.

For several years, members of our two divisions continued to meet to discuss the psychology of both war and peace. Friendships were formed and many Division 19 members joined Division 48, and vice versa. In many ways, I believe that we should merge into a single division. We could pool our resources and perhaps be a more vocal and effective voice for the advancement of peace, as well as the ethical use of psychological knowledge in war.

It is interesting that our society pays so much more attention to war than to peace. As a citizen, it seems to me that our overriding tendency is to use war to resolve conflicts with other nations or ideologies. As I write, a revolution is occurring in Syria. Many innocent people are being killed or displaced. There are many politicians—but I can assure you almost no soldiers—who call for the use of US armed forces in helping the rebels achieve their mission. Maybe this is another case of America's preference for "instant pudding"—that is, a dramatic and fast short-term solution over that of a more difficult long-term political solution that might take years to achieve.

It is also fair to question why peace psychology receives so little attention among the ranks of psychologists, in the press, and in society. For one thing, psychologists follow the money. The defense industry provides hundreds of millions of dollars each year to fund research that incorporates military psychology. Military laboratories, Department of Defense contractors, and grants to university professors support a dizzying array of basic and applied psychological research aimed at improving soldier performance and military effectiveness. And, of course, huge sums of money have been allocated to providing mental health services to military members, especially in this era of prolonged war.

Where is the funding for peace psychology? What if the Pentagon was not simply the Department of Defense (in effect, the Department of War), but rather the Department of War and Peace? What if the government funded a "peace machine" to the same extent it funds the war machine?

There would be a cornucopia of dividends in such a world. At a practical level, peace is a lot less expensive than war. Solving problems through diplomacy and peaceful policies would allow a huge decrease in the defense budget. A peace-oriented society would still need to train and maintain an effective military to deter aggression, but I suspect that few nations would consider attacking a nation armed with the military technologies possessed by the United States. A fair and balanced foreign policy based at least in part on sound psychological principles could reduce the motive for other nations or groups to attack.

I think that psychology has many concrete suggestions to offer in terms of promoting peace. To fully cover these would require another book. But here is a quick review of some things that psychologists have learned about promoting peaceful behavior, decreasing conflict, and building cooperation.

Human Mindsets Can Be Changed

Psychologist Carol Dweck received the APA Award for Distinguished Scientific Contributions in 2011 for her research into how mindsets can affect social attitudes and behavior. A mindset is a framework through which a person views the world.

Dweck distinguishes between a "fixed" mindset and a "growth" mindset. People with a fixed mindset believe that human traits such as personality, social beliefs, and intelligence are unchangeable. People with a growth mindset look at others as potentially able to show transformational changes in these traits.

It turns out that the type of mindset one has about others predicts outcomes that are relevant to the psychology of peace. Dweck reviews research showing that Jewish Israelis with fixed mindsets had more negative attitudes toward Palestinians than those who had growth mindsets.[4] Moreover, researchers showed that mindsets could be manipulated through experimental interventions. When exposed to a treatment that increased the growth mindset participants became more likely to express favorable beliefs about Palestinians and greater openness to make compromises to allow peace. These effects were not limited to Jewish Israelis. Similar studies done with Palestinian Israelis showed parallel results. Those in whom a growth mindset was induced were more amenable to negotiation and compromise than those with a fixed mindset.

Findings like these are tremendously important, and this and similar research should be funded with the same sense of necessity as military psychology. It is plausible that educational curricula could be developed that would more effectively nurture a growth mindset in children. If this mindset persevered, a generation of Jewish and Palestinian Israelis would someday come to political power that would be less intractable and more open to reasonable compromise than those of the current generation.

Using psychology-based methods to encourage and engender beliefs congruent with a growth mindset could achieve great success if adopted across the world. It is much easier to bomb an enemy than it is to change the way a population thinks. Military action may stop a foe from attacking, but it may also cause lasting anger and hostility against the victorious nation. Changing mindsets would take much longer to achieve but I suspect the results would be more durable, and would also set the occasion for peaceful cooperation, not just the absence of war.

Improving Quality of Life

Biological psychologists know that malnutrition among pregnant women results in a host of problems for the child as he or she matures. High stress affects the development of the fetal brain. Psychoactive drugs, both licit and illicit, have devastating effects on brain development. There is evidence that children who experienced these prenatal stressors are more prone to violence and aggression as they mature.[5]

So part of a comprehensive peace psychology agenda would be to ensure that mothers receive adequate nutrition while pregnant and learn avoid certain drugs. Most of you know about fetal alcohol syndrome, where babies born to mothers who drink alcohol are subject to a variety of neurological and behavioral deficits. Less well understood, but equally dangerous, are the effects of nicotine, cocaine, and many other powerful drugs on the developing fetus. Education and prenatal care will never completely eliminate these abuses, but it will help a great deal.

High levels of stress in mothers also impact the neurological and psychological development of the child.[6] You can't prevent stress, but through prenatal education mothers can be taught the resilience skills needed to cope effectively with the challenges of life. If we can use Comprehensive Soldier Fitness to help soldiers deal with stress, then we can use similar procedures to give expectant mothers the same skills. This will give the fetus, and subsequently the child, a positive edge. It is much better to begin life with a healthy brain than one compromised by poor nutrition, drugs, or excessive stress hormones.

Common Goals Can Trump Group Differences

In a classic social psychological experiment, Muzafer Sherif studied the behavior of boys in summer camp.[7] Boys were randomly assigned to a given group, given separate living quarters and a unique name ("eagles" or "rattlers"), and were pitted against each other in various games. They soon became very competitive and sometimes aggressive toward each other. Forming these groups and giving them unique identities, in the context of competition, set the occasion for sometimes dysfunctional behaviors.

Sherif then brought the groups together in different ways to see how competition and aggression could be decreased. Interestingly, having the two groups engage in an enjoyable but passive activity like watching a movie not only did not help, it actually increased the hostility between groups. Sherif then contrived situations where the boys had to work together to achieve difficult goals, such as pulling together on a rope to get a bus started. Working together on such tasks was effective in reducing hostility and tension.

Such an approach can be used between nations. What if Israelis and Palestinians had a larger, supraordinate goal that neither group could achieve independently but could be accomplished by working together? For instance, it appears that neither group can attain peace by themselves. If they could come to view peace as the common goal and work together to achieve it, perhaps the necessary compromises and concessions needed could be granted by each side.

Based on this principle, what if the United States, Russia, China, and other nations decided to pool their resources to achieve a truly significant scientific achievement that no one country could do alone? Sending humans to Mars might be an example. Sharing in the funding, basic science, engineering, and training needed to accomplish this mission would almost certainly allow these nations to work together in a more cooperative manner on matters of disagreement. Sending humans to Mars would most likely be a lot less expensive than fighting a war among these nations.

Political Behavior May Be Operantly Conditioned

B. F. Skinner extolled the power of reinforcement in changing human behavior. Sometimes I think we have forgotten those lessons. Parents can be taught to reward cooperative behavior and nonaggressive play. They can model both verbally and through their actions the importance of nonviolent solutions to conflict. These approaches can be systematically incorporated into the school system. In the study of history, nations that achieve greatness through peace should be celebrated more than those who turn to war to solve their problems.

Operant conditioning principles were developed to apply to individual organisms, but it is plausible that they could be systematically applied to shape and modify the behavior of aggregated groups, such as nations or ideologically aligned groups. Principles of administering rewards and punishments contingent on a targeted group displaying certain behaviors could greatly impact their international behavior.

One of the core findings in operant conditioning is that reward is more effective than punishment in effecting long-term behavioral change. Punishment is relatively easy to apply and gets the attention of the target but has a host of undesirable side effects.[8] For example, spanking a child will likely result in a temporary cessation of the troublesome behavior. But it is problematic on several levels: (1) Punishment by itself does not tell the child what behavior he or she *should* display in that situation; (2) the use of physical punishment models interpersonal aggression as a solution to life's problems and the child may in turn come to rely on aggression as a prime option in dealing with his or her own interpersonal difficulties; (3) physical punishment can result in emotional trauma for the child; (4) unless it is extremely violent, punishment tends to simply suppress the unwanted behavior, not eliminate it; and (5) painful punishment can elicit what psychologists call "respondent" aggression, that is, the child may lash out in pain at the punisher or others nearby.

A foreign policy based on a punishment model is similar in its effects on the targeted nation. A military strike designed to punish an errant nation is in many ways easier to plan and execute than a long-term effort to reward appropriate behavior. In and of itself, military force suffers from the same drawbacks as the use of punishment to alter individual behavior. At the very least, it further engenders an international climate of violence and thus tends to perpetuate war as the primary mode for solving international political problems.

Changing behavior through reward, on the other hand, requires greater thought and consistency and may take longer to achieve desired results. Anyone who has ever shaped a rat to press a lever to obtain food in a Skinner box knows that patience is a virtue.[9] The challenge for a foreign policy that employs the operant model in its dealings with other nations is similar. Despite changing elected national leadership and senior political appointments, a reward-based policy must be systematic and consistent. Properly implemented, however, such an approach should yield more positive and enduring change than a foreign policy based on the application of military force, embargoes, blockades, and the like.

Social Psychologists Have Many Tools for Furthering Peaceful Behavior

Many of the attitudes and behaviors that contribute to violence and a warlike society have been the subject of extensive study by social psychologists. From Stanley Milgram's classic work on obedience to authority to contemporary research such as that described above by Carol Dweck, we know a lot about the conditions that promote conflict, prejudice, and aggression.

Here are just a few topics studied by social psychologists that are relevant to nurturing nonviolence and peace versus aggression and war:

- *In-group, out-group bias.* This is the tendency for groups to adopt and share common attitudes among their members but to be prejudiced against people outside of their group. Social psychologists tell us there are various strategies for reducing this bias.
- *Attribution bias.* This is the tendency to explain your own behavior according to situational factors but to explain the behavior of others according to internal causes, and is similar to another phenomenon known as the *fundamental attribution error.* At the group level this means showing greater sensitivity and understanding to actions taken by your nation but believing that other nations behave in some undesirable ways because of stable, internal causes, such as hostility toward your country. Politicians calling other nations "evil" are more than name-calling. They imply a perverse and unchanging internal cause of that nation's political positions. At the individual level these attributions are often wrong. When they are wrong at the national level, this may contribute to poor relations between the nations involved.
- *The self-serving bias.* At the individual level this means being very forgiving of your own faults and misdeeds, attributing them to external causes, but taking full credit for your personal accomplishments. Similar attributions can be made at the group or collective level.
- *Prejudice and discrimination.* Factors that contribute to the formation of prejudice and discrimination as well as factors that serve to diminish it are well known.

Something known as the mere exposure effect may, for example, result in a greater understanding of others, as well as decreasing prejudicial behavior.[10] During the Cold War there was little travel between the United States and its chief threats, the Soviet Union and China. It is no accident that tensions began to ease as contact—both at the level of heads of state and among individual citizens—became commonplace.

- *The psychology of stereotyping and people perception.* Social psychologists have learned that we quickly form schemas in order to make hasty judgments about others. A schema is a mental shortcut, often based on superficial information that we use to categorize others. Being told that a woman you meet at a party is a college professor may elicit a completely different evaluation of her compared to if you were told she was an actress. Collectively we use schemas to categorize other nations and groups. Many Americans hold a dim view of Islamic countries, based on widely publicized incidents involving Islamic countries that are not, in fact, representative of most such nations.

National Education Policy

Education may provide a key approach to reducing violence and promoting peaceful behavior. A curriculum of multicultural education beginning at the earliest grade levels and continuing through higher education would provide future generations, including individuals who emerge as policy makers, with a frame of reference that would be more accepting of the diverse world of the twenty-first century. Psychologists have the knowledge and tools to help schools implement these programs. It will take time to have an impact but it will be worth the wait.

Base International Relations on Positive Psychology Principles

Positive psychology brings a scientific understanding of all that is excellent about human beings. Resilience, positive character, personal growth, and flourishing are prime concepts. A nation whose leaders develop policies that nurture this growth in its citizens, and that also uses this as a frame of reference for its dealings with other nations would be more likely to support peace than to turn to war. Christopher Peterson, a pioneer of positive psychology, often said that other people matter. Maybe it is time to have a national policy that truly adheres to an extension of this thought: "other nations matter."

ETHICS AND MILITARY PSYCHOLOGY

Philosophers, theologians, political and social scientists, and psychologists have long struggled over the question of whether war and aggression is a core feature of human nature or something that emerges from dynamic behavioral, social, and political processes. In other words, is war inevitable by virtue of being human, or is it something that can be prevented by behavioral and social engineering?

I certainly do not know the answer to this question. But as a pragmatist, I expect war to continue to be part of our social reality through the foreseeable future. The world is simply too fragmented along political, social, economic, and religious lines to set the occasion for peaceful cooperation among nations. More disturbing, in many respects, is the emergence of ideological warfare exemplified by the events of September 11, 2001, and many similar events around the world, both before and after that date. Eisenhower may have been right about some people wanting peace, but there will always be individuals, groups, or nations willing to inflict armed force on others.

It is a safe bet that the United States and most other countries must continue to prepare for war through the twenty-first century. As we have seen in this book, psychologists will play a critical role in preparing the military for war, aiding it in the execution of war, and in recovering from its aftereffects. And as psychology plays out its role with the military, inevitably ethical questions must arise about its role in the process.

People become psychologists for many reasons but a common theme is to "do good" for others. To promote psychological health, treat psychopathology, aid the elderly, improve productivity, and educate our youth—all of these are noble pursuits. As psychologists we share core beliefs. We believe that human behavior is the product of natural forces and that science provides the best lens for improving our understanding of it. Psychologists believe that human rights are of paramount importance and many psychologists actively fight for these inalienable rights, both domestically and abroad. We believe that all people have the right to a safe and productive life, and that people should be judged on their individual merits, not by the color of their skin, their religion, or who they develop intimate relationships with.

War presents an ethical dilemma to psychologists. On the one hand there have been times throughout history when powerful enemy states have threatened these core values. Even for peace loving professionals, turning a blind eye in the 1930s and 1940s to the atrocities perpetrated by Germany against Jews and other targeted groups, and Japan's atrocious and unrestrained violence in China and elsewhere in Asia was not an option. A Europe dominated by the Third Reich or the Pacific ruled by the Japanese would have led to the subjugation and death of millions of human beings. Something had to be done, and psychologists scrambled to do their part.

Since World War II things have not been so black and white. The United States went to war in Korea and Vietnam for reasons that were not made clear to its citizens. The domino theory notwithstanding, I don't think many people believed that a victorious North Vietnam would soon be landing soldiers on Malibu beach. With the advent of widespread and daily televised news coverage, for the first time in history ordinary citizens could witness the cruel brutality of war in their own homes. For those of us who remember the Vietnam war, who can forget the image of the young girl fleeing a napalm strike, her clothes burned off and skin hanging from her burns, or the weekly reports of the American dead?

The wars of the twenty-first century may seem even more perplexing. Despite the gravity of the attacks of September 11, 2001, they were not perpetrated by a nation-state. Imagine how our sense of national purpose would have been much clearer if it had been Iranian warplanes that destroyed the World Trade Center and attacked the Pentagon. The citizens of the United States and those of many countries around the world would have rallied against a clear foe. A military reprisal might not have even been necessary if the nations of the world came together to condemn such an act of unprovoked aggression. And if military force was necessary, clear objectives could be identified, and the conditions of victory defined.

Instead the United States was left with what former Secretary of Defense Robert Gates referred to as "wars of choice." Following December 7, 1941, the United States had to wage war against the Axis powers. In contrast, after September 11, 2001, there were many other options on the table. Politicians and political scientists can say what they wish, but at the end of the day there was not a clearly compelling rationale for invading Iraq or for widespread military action in Afghanistan. Indeed, it is commonly argued that the war in Iraq diverted strategic attention away from the effort to defeat Al Qaeda and to bring Osama bin Laden to justice. One must wonder how the economic and political status of the United States would have been different if its elected political leaders had stayed focused on defeating those who actually attacked the United States and had not engaged in general warfare in Afghanistan and Iraq. Regardless of the economic and political consequences of these wars, I can personally think of several former cadets and colleagues who would still be alive today, along with many thousands of other US troops and countless thousands of innocent citizens of Afghanistan and Iraq, if a more peaceful solution to our nation's problems had been pursued.

In addition to the questionable rationales for these wars, the US military found itself in a series of widely publicized criminal actions against Afghans and Iraqis. As a result of the situation at the now infamous Abu Ghraib military prison, the allegations of the torturing of suspected terrorists, and the actions of rogue soldiers like Sergeant Robert Bales. who stands accused of murdering 16 Afghan citizens, many psychologists experienced a conflict between their core beliefs and the actions of the military (at least as commonly portrayed in the media).

By the summer of 2007, as I began my year-long term as the president of the Society for Military Psychology (Division 19 of the American Psychological Association), outrage over these wars was peaking. During the annual APA convention in San Francisco that August there were highly visible protests, especially aimed at allegations that some psychologists had participated in torturing prisoners of war.

The anger was sometimes expressed very personally. On one occasion I witnessed a protester—and to be fair I could not say for sure if he was a psychologist or not—attempt to physically attack a uniformed military psychologist as he was walking along a downtown street.

In some cases, this anger was turned toward any psychologist who was perceived as promilitary. Former APA president Martin Seligman was accused by some of assisting the military in developing torture techniques by reverse-engineering his concept of learned helplessness.[11] Fortunately, personal attacks of this sort were not widespread and seemed to come from only a small minority of psychologists.

The antimilitary sentiment also found expression against programs aimed to improve the adjustment and resilience of military members. The Army's Comprehensive Soldier Fitness (CSF) program was the target of much criticism.[12] Following the publication of a special issue of the *American Psychologist* in January 2011 that described CSF, a number of psychologists objected to it for a variety of reasons. It is interesting that a program like CSF, with its potential to help the lives of literally millions of military members, their families, and Department of Defense civilians, would be so vocally opposed by—once again—a tiny minority of psychologists. It is even more disturbing since the majority of the criticism seems based on political orientation rather than a critique of the psychological foundation of the program.

I am not criticizing psychologists who take a dim view of psychologists working on behalf of the military. Their objections can be weighed by the community of psychologists and can help guide the further development of programs like CSF both in terms of improving effectiveness, and also at an ethical level. We owe the military effective *and* ethical psychologically based interventions and programs. A spirited discussion on the part of psychologists with disparate views of the role of psychology and the military contribute to this goal.

It is interesting, upon reflection, to compare the roles of physicians and psychologists in the military. It is almost unheard of to hear someone declare that physicians should not work on behalf of the military. Military physicians offer medical care to soldiers and their families. In times of war they are present at military hospitals where wounded enemy soldiers and civilian noncombatants are also treated, at least under some circumstances. We did not read much about it in the press, but military physicians were present at Abu Ghraib as well as Guantanamo Bay military prisons. I don't recall anyone claiming that these physicians were violating human rights.

However, the role of clinical psychologists at Abu Ghraib and Guantanamo Bay was vehemently criticized, mostly by other psychologists. Some expressed the belief that clinical psychologists should be banned from providing treatment to enemy prisoners of war. One can scarcely imagine physicians criticizing military physicians who offered medical treatment to enemy POWs in these settings.

I believe that the ambiguous nature of twenty-first century warfare is the root cause for this disagreement about the role of psychologists in war. There is no clear and easily defined enemy, only an international network of avowed terrorists. Under these conditions, it is hard for the nation's elected leadership to identify the nature of the threat and to rally the population, including psychologists, against that threat. The enemy appears and then is absorbed back into his neighborhood, town, and nation. His motives may be ideological rather than strictly political. The laws of war that were based on traditional conflicts between warring nations in the twenty-first century do not apply to the sorts of conflicts experienced thus far in this century and those likely to occur over the next 25 years or more. If taken prisoner, it is not clear what rights the enemy combatant may have. And of course more traditional threats to our national security may also emerge, which further complicates the state of affairs.

American Psychological Association Stance

It is in this murky context that the American Psychological Association began to reexamine its ethical standards as they relate to the role of military psychologists, particularly in the treatment of enemy prisoners of war. What professional roles are appropriate and which ones are not?

Most of the controversy surrounds the topics of human rights and interrogations. There were reports in the press of psychologists assisting the military and the Central Intelligence Agency (CIA) in torturing prisoners. From reading these articles, you could be excused for concluding that psychologists were heavily involved in this activity. Nothing could be further from the truth. I know a good number of military psychologists who have been assigned to duty at Guantanamo prison and similar locations, and in no single case have they observed a psychologist assisting in unethical acts, let alone doing so themselves. I also have had the occasion to speak with operator/agents from other government agencies, and they confirmed that no psychologists were involved in even the controversial but legal (at least by the standards of the Bush administration) "enhanced" interrogation techniques. My best guess is that the agents/interrogators used psychological principles easily obtained from introductory psychology books, but that there were no (or at worst very few and isolated) instances of psychologists being actively involved.

Regardless, the question of what constitutes ethical behavior for psychologists who are employed by the military is a very important question. Zimbardo's infamous Stanford Prison experiment demonstrated that ordinary people may be overcome by the demand characteristics of a given situation and behave in unethical, illegal, and potentially dangerous ways.[13] Given the critical role of psychology in twenty-first century warfare, it is essential that psychologists have a clear code of conduct available to guide their work in support of national defense.

Psychologist Stephen Behnke, chief ethics officer of the APA, recently reviewed the history of the organization's formal stance on the role of psychologists in military situations where human rights are of paramount importance.[14] Between 2004 and 2010 the APA revised and revised, again and again, its formal policy regarding ethics and interrogation.

Behnke points out that the APA Ethics Code had for many years been clear on most ethical issues faced by psychologists, including the treatment of experimental subjects, authorship, and so forth. But the code was far less explicit in defining ethical conduct in the national security arena in which, with the wars in Iraq and Afghanistan, military psychologists were finding themselves. Accordingly, in 2004 Ron Levant, APA president at the time, established the Presidential Task Force on Psychological Ethics and National Security (PENS). While PENS was tasked to look at national security-related issues in general, most of its attention ended up focusing on the role of psychologists in interrogations. This distinction is important and I will return to it shortly.

By 2005, the PENS task force had generated 12 statements addressing the ethical conduct of psychologists working in situations where interrogations occur. These 12 statements all relate to core ethical beliefs held by psychologists. These include the notion that psychologists do not inflict harm on others, that their ethical obligations extend to all people involved in a given context, and that they do not go beyond their professional competencies. A full listing of the 12 principles is easily available at the APA website (www.apa.org).[15]

Over the next several years a number of resolutions further refined and clarified the APA Ethics Code in this matter. This iterative process represents an interesting case study of how complex ethical standards evolve, being stretched one way or another by input from concerned constituents. These resolutions fine-tuned various aspects of the original 12 principles. A resolution issued in 2006 provided a more international perspective on the issue by providing explicit reference to the stance of the United Nations on torture, and it updated the definition of torture to be in line with that of the United Nations. A resolution in 2007 more clearly stated that there are no circumstances wherein torture may be tolerated and provided specific examples of techniques, such as waterboarding, that exemplify torture. It also called upon the US government to ban torture and for the US courts system to reject testimony obtained through such techniques. Yet another amendment issued in 2008 further reinforced, using stronger and more explicit language, the APA stance against torture, the responsibilities of psychologists during interrogations, and the prohibition against psychologists engaging, aiding, or abetting in torture.

All told, the APA revision of the Ethics Code recognizes the valuable contributions that psychologists can make in affairs relating to national security, and also helped psychologists employed in these settings better understand the ethical boundaries they face. Although the language of the PENS report and the subsequent resolutions are legalistic and somewhat arcane, they can be summarized in a simple, very reasonable statement: Psychologists will not engage in torture or tolerate that behavior in others. Not all psychologists will agree on what defines torture and exactly what role a psychologist may assume in interrogations or, more broadly, in affairs of national security. It was important for the APA to publicly state, in its official policy, strong opposition to torture. In the end it is the responsibility of all psychologists to reconcile their actions with both APA policy and their own professional ethos.

Ethical Considerations in Other Military Psychology Contexts

Although the PENS report purports to address the role of psychologists in matters of national security in general, in fact it focused almost exclusively on the ethics of interrogation and torture. But as you have seen, the vast majority of military psychologists engage in work that is unrelated to this issue. Are there ethical issues in these other contexts?

Suppose a human factors psychologist is part of the design team for a semiautonomous airborne weapons system that has the capability to acquire a target and then decide, without direct human intervention, to engage and destroy the target. Once deployed, further suppose that this system identifies a high-value target, perhaps a terrorist leader. It ultimately makes the decision to kill the target, but in the process several innocent people are also killed. Does the individual psychologist who helped design the system shoulder a responsibility for the death of these innocent civilians? Should the APA have an official position on such a circumstance?

Almost everything that psychologists do in support of the military is ultimately aimed at creating a more lethal and effective organization. The mission of the air force is to fly and fight; achieving this mission will inevitably result in the loss of life, often innocent ones. As an extreme example, does a psychologist who helps select and screen personnel to operate systems armed with nuclear weapons violate the ethical code of the APA, or his or her own code of ethics, if someday the personnel who he or she screened and selected launch a missile that kills hundreds of thousands of human beings?

In my own role as a professor at West Point my fundamental mission is to help educate, train, and inspire leaders of character who may lead soldiers into combat. Combat is a dirty business, and doubtlessly officers I have helped educate have been involved in actions that have not only killed enemy combatants, but also men, women, and children who had not taken up arms against the United States. There has never been a war where only soldiers were killed. Collateral damage is unavoidable.

And what about the clinical psychologist who offers therapy to military personnel? The goal is not just to cure a pathology, but also to enable the soldier to return to his or her unit and rejoin the fight. One might argue that without psychological treatment offered by the military clinical psychologist, these soldiers would never have been able to return to duty. But those who respond to treatment and return to duty sometimes kill others. Again, does this violate the ethical principles that define what psychology is all about?

The Role of Military Psychologists is Both Ethical and Vital to National Defense

If there was a clear and easy answer to these questions I would not need to write this chapter. There are many mitigating circumstances that apply. But the consensus seems to be that in a lawful war it is within the legitimate scope of duties for psychologists to leverage their knowledge to help the military prepare, fight, and recover from war. There are clearly some circumstances—interrogations and torture being the prime example—where even within the context of lawful war psychologists must not engage in certain activities.

There will always be those who maintain that any association of psychologists with the military is unethical. But the threats to our country are real, and psychologists, more so than ever before, have the expertise to vastly improve the effectiveness of our military forces. Most psychologists concur. Since World War I they have rushed forward to help our military when it needed them the most.

FINAL THOUGHTS

This has been a book about how psychology can help win the wars of the twenty-first century. We have learned through hard experience that firepower alone will not win contemporary wars. The days of two (or more) countries slugging it out with rifles, rockets, and bombs until one side surrenders seem to be gone, and let's all hope that such a war never occurs again. The power of modern weapons would likely leave the earth a smoldering heap of rubble. The technological and political realities of twenty-first-century warfare have catapulted psychology to the status of a major difference maker in the outcome of current and future conflict.

I hope that world leaders will eventually come to capitalize on psychology to promote peace instead of war, with the same degree of urgency and level of funding that they have shown in using psychology to improve military effectiveness. Today, unfortunately, the world remains a very dangerous place. Large democratic nations are the targets of various ideological extremists and totalitarian countries. Many smaller nations, especially in parts of the Middle East and Africa, are fighting prolonged wars against insurgents. These wars have the potential of becoming regional wars without timely and thoughtful reactions by other nations.

The "nuclear club" is growing. Iran and North Korea may be close to producing powerful nuclear weapons and both have the missiles needed to unleash them on other countries. A nuclear weapon or even a so-called dirty bomb may soon fall into the hands of terrorists. Because terrorist groups are not formally aligned with a given nation, they may be able to detonate such a device with little concern of retaliation.

Cyber warfare is now a distinct possibility. A computer attack could shut down the nation's power supply, disable satellite communication, and disrupt the financial system. You saw how terrible these effects were following Hurricane Sandy. During the storm and for a few days following it, I had no power, phone service, or cable television. This disconnection from the world was disturbing, and I could not help thinking what it would be like if the entire country suddenly found itself in the same situation. Cyber warfare is no laughing matter.

So I end this book with a hope for a more peaceful world, one where psychologists can employ their talents to improve well-being, and not focus on the preparation, execution, and recovery from war. But until that day comes psychologists must continue to support the military. Former APA president and founding father of positive psychology Martin Seligman said it best:

> *Three ideologies have arisen in the past century that have sought to overthrow democracy by force: fascism, communism, and jihadist Islam. It should be noted that without a strong military and the will to use force responsibly in our self-defense, our victories would not have happened, and defense against current and future threats would be impossible. Psychology materially aided in the defeat of the first two threats, and in doing so it carved out its identity. We are proud to aid our military in defending and protecting our nation right now, and we will be proud to help our soldiers and their families into the peace that will follow.*[16]

NOTES

1. From a radio and television broadcast with British Prime Minister Harold Macmillan in London, August 31, 1959; archived at The American Presidency Project, www.presidency. ucsb.edu/index.php (the specific quote is at www.presidency.ucsb.edu/ws/index.php?pid= 11485&st=Peace&st1=).

2. B. F. Skinner, *Beyond Freedom and Dignity.*

3. Janice H. Laurence and Michael D. Matthews (eds.), *The Oxford Handbook of Military Psychology*.

4. For a good review of Dr. Dweck's groundbreaking work, see Carol S. Dweck, "Mindsets and Human Nature," 614–22.

5. For example, see Tanya M. M. Button, Barbara Maughan, and Peter McGuffin, "The Relationship Between Maternal Smoking to Psychological Problems in the Offspring," 727–32.

6. For example, there is evidence that high maternal stress is associated with the subsequent development of schizophrenia in their offspring, many years following birth. See Ali S. Khashau et al., "Higher Risk of Offspring Schizophrenia Following Antenatal Maternal Exposure to Severe Adverse Life Events," 146–52.

7. Muzafer Sherif et al, *The Robbers Cave Experiment*.

8. The adverse side effects of using punishment to alter behavior have been known to psychologists for nearly 50 years. For an excellent discussion see Nathan H. Azrin and William C. Holz, "Punishment," 380–447.

9. Designed by psychologist B. F. Skinner and used in many of his experiments, this box is used to administer rewards and punishments to animals as they learn various tasks.

10. The classic paper on the mere exposure effect was published in 1968. It stimulated hundreds of subsequent studies on this social psychological process. See Robert B. Zajonc, "Attitudinal Effects of Mere Exposure," 1–27.

11. For a stirring rebuttal to these claims, see Dr. Seligman's response: "A Letter to the Editor by Martin Seligman," *VoltaireNet.org*, June 20, 2010, www.voltairenet.org/article165964.html.

12. For example, see Roy Eidelson and Stephen Soldz, "Does Comprehensive Soldier Fitness Work? CSF Research Fails the Test," *Coalition for an Ethical Psychology*, May 2012, www.ethicalpsychology.org/Eidelson-&-Soldz-CSF_Research_Fails_the_Test.pdf.

13. Zimbardo reviews this classic experiment and offers a compelling update on its relevance to many of the sociopolitical issues facing contemporary society in his book, Philip Zimbardo, *The Lucifer Effect*.

14. Stephen H. Behnke and Olivia Moorehead-Slaughter, "Ethics, Human Rights, and Interrogations," 50–62.

15. You can search for "ethics" on the home page or go directly to the ethics statement with the following link: www.apa.org/ethics/code/index.aspx.

16. Martin E. P. Seligman and Raymond D. Fowler, "Comprehensive Soldier Fitness and the Future of Psychology," 86.

BIBLIOGRAPHY

Adorno, Theodor W., Else Frenkel-Brunswik, Daniel J. Levinson, and R. Nevitt Sanford, *The Authoritarian Personality* (New York: Harper and Row, 1950).

Aldwin, Carolyn M., Michael R. Levenson, and Avron Spiro, "Vulnerability and Resilience to Combat Exposure: Can Stress Have Positive Effects?" *Psychology and Aging* 9:1 (1994): 34–44, doi:10.1037/0882-7974.9.1.34.

Algoe, Sara B., and Barbara L. Fredrickson, "Emotional Fitness and the Movement of Affective Science from Lab to Field," *American Psychologist* 66:1 (2011): 35–42, doi:10.1037/a0021720.

Alpert, Jon, Ellen Goosenberg Kent, and Matthew O'Neill, *Wartorn 1861–2010* (New York: HBO Documentary, 2010), DVD.

American Psychiatric Association, *Diagnostic and Statistical Manual of Mental Disorders DSM-IV-TR*, 4th ed., 463–68 (Arlington, VA: American Psychiatric Publishing, Inc., 2000).

Andersen, Richard A., Eun Jung Hwang, and Grant H. Mulliken, "Cognitive Neural Prosthetics," *Annual Review of Psychology* 61 (2010): 169–90, doi:0.1146/annurev.psych.093008.100503.

Andrews, Lincoln C., *Military Manpower: Psychology Applied to the Training of Men and the Increase of their Effectiveness* (New York: Dutton, 1920).

Azrin, Nathan H., and William C. Holz. "Punishment," in *Operant Behavior: Areas of Research and Application*, ed. Werner K. Honig (New York: Prentice-Hall, 1966), 380–447.

Barbara Kingsolver, *Animal Dreams* (New York: Harper Perennial, 1991).

Bartone, Paul T., "The Need for Positive Meaning in Military Operations: Reflections on Abu Ghraib," *Military Psychology* 17:4 (2005): 315–24, doi:10.1207/s15327876mp1704_5.

Bartone, Paul T., Dennis R. Kelly, and Michael D. Matthews, "Psychological Hardiness Predicts Adaptability in Military Leaders," *International Journal of Selection and Assessment* 21:2 (2013): 200–10, doi:10.1111/ijsa.12029.

Behnke, Stephen H., and Olivia Moorehead-Slaughter, "Ethics, Human Rights, and Interrogations: The Position of the American Psychological Association," in *The Oxford Handbook of Military Psychology*, eds. Janice H. Laurence and Michael D. Matthews (New York: Oxford University Press, 2012), 50–62.

Belenky, Greg, David Penetar, David Thorne, Kathryn Popp, John Leu, Maria Thomas, Helen Sing, Thomas Balkin, Nancy Wesensten, and Daniel Redmond, "The Effects of Sleep Deprivation on Performance During Continuous Combat Operations," in *Food Components to Enhance Performance*, ed. Bernadette M. Marriott (Washington, DC: National Academy Press, 1994), 127–35.

Beyerchen, Alan D., "Clausewitz, Nonlinearity, and the Importance of Imagery," *International Security* 17:3 (Winter, 1992): 59–90.

Boller, Paul F. Jr. and John H. George, *They Never Said It* (New York: Oxford University Press, 1989), 124.

Boring, Edwin G., *A History of Experimental Psychology*, 2nd ed. (New York: Prentice-Hall, 1950).

Brands, H. W., *The First American: The Life and Times of Benjamin Franklin* (New York: Doubleday, 2000), 620.

Brokaw, Tom, *The Greatest Generation* (New York: Random House, 1998).

Brown, Curtis, ed., *Roots of Strategy, Book 2: 3 Military Classics* (Mechanicsburg, PA: Stackpole Books, 1987).

Button, Tanya M. M., Barbara Maughan, and Peter McGuffin, "The Relationship Between Maternal Smoking to Psychological Problems in the Offspring," *Early Human Development* 83:11 (2007): 727–32, doi:10.1016/j.earlhumdev.2007.07.006.

Cacioppo, John T., Harry T. Reis, and Alex J. Zautra, "Social Resilience: The Value of Social Fitness with an Application to the Military," *American Psychologist* 66:1 (2011): 43–51, doi:10.1037/a0021419.

Calaprice, Alice, *The New Quotable Einstein* (Princeton, NJ: Princeton University Press, 2005), 173.

Campbell, Donald J., "Leadership in Dangerous Contexts: A Team-Focused, Replenishment-of-Resources Approach," in *The Oxford Handbook of Military Psychology*, eds. Janice H. Laurence and Michael D. Matthews (New York: Oxford University Press, 2012), 158–75.

Chamberlain, Joshua Lawrence, *The Passing of Armies: An Account Of The Final Campaign Of The Army Of The Potomac* (New York: G. P. Putnam's Sons, 1915), 295.

Chatham, Ralph E., *A Possible Future for Military Training* (Arlington, VA: DARPA, 2003), https://campus.georgetown.edu/@@/23BC4B7810D027E5CE92F7A587E2CE63/courses/1/SEST-562-01.Fall2011/content/_2089300_1/DARPA_TrainingVision_5D8BF.pdf.

Christeson, William, Amy Dawson Taggart, and Soren Messner-Zidell, *Ready, Willing, and Unable to Serve: 75 Percent of Young Adults Cannot Join the Military* (Washington, DC: Mission: Readiness, Military Leaders for Kids, 2009).

Coram, Robert, *Boyd: The Fighter Pilot Who Changed the Art of War* (New York: Little, Brown, 2002).

Cornum, Rhonda, and Peter Copeland, *She Went to War: The Rhonda Cornum Story* (Novato, CA: Presidio Press, 1992).

De Angelis, Karin, and David R. Segal, "Minorities in the Military," in *The Oxford Handbook of Military Psychology*, eds. Janice H. Laurence and Michael D. Matthews (New York: Oxford University Press, 2012), 325–43.

DeGroot, Adriaan, *Thought and Choice in Chess* (The Hague: Mouton, 1965). (Original work in Dutch published in 1946.)

Dewey, Larry, *War and Redemption: Treatment and Recovery in Combat-Related Traumatic Stress Disorder* (Aldershot, UK: Ashgate, 2004).

Duckworth, Angela L., Christopher Peterson, Michael D. Matthews, and Dennis R. Kelly, "Grit: Perseverance and Passion for Long Term Goals," *Journal of Personality and Social Psychology* 92:6 (2007): 1087–101, doi:10.1037/0022-3514.92.6.1087.

Durkheim, Emile, *Suicide: A Study in Sociology* (Glencoe, IL: Free Press, 1951).

Dweck, Carol S. "Mindsets and Human Nature: Promoting Change in the Middle East, the Schoolyard, the Racial Divide, and Will Power," *American Psychologist* 67:8 (2012): 614–22, doi:10.1037/a0029783.

Dyer, Jean L., and Geoffrey H. Martin, *The Computer Background of Infantrymen: FY99, Research Report 1751* (Arlington, VA: US Army Research Institute for the Behavioral and Social Sciences, 1999).

Dyer, Jean L., John Reeves, and Richard L. Wampler, *Training Effectiveness Analysis (TEA) of the Land Warrior (LW) System: Phase I — The Baseline Platoon* (Ft. Benning, GA: US Army Research Institute for the Behavioral and Social Sciences, 1998).

Dyer, Jean L., and Jennifer S. Tucker, *Training Analyses Supporting the Land Warrior and Ground Soldier Systems, Research Report 1904* (Alexandria, VA: US Army Research Institute for the Behavioral and Social Sciences, 2009).

Endsley, Mica R., "Toward a Theory of Situation Awareness in Dynamic Systems," *Human Factors* 37:1 (1995): 32–64, doi:10.1518/001872095779049543.

Endsley, Mica R., Leonard D. Holder, Bruce C. Leibrecht, Daniel J. Garland, Richard L. Wampler, and Michael D. Matthews, *Modeling and Measuring Situation Awareness in the Infantry Operational Environment, Research Report 1753* (Alexandria, VA: US Army Research Institute for the Behavioral and Social Sciences, 2000).

Feder, Adriana, Steven M. Southwick, Raymond R. Goetz, Yanping Wang, Angelique Alonso, Bruce W. Smith, Katherine R. Buchholz et al., "Posttraumatic Growth in Former Vietnam Prisoners of War," *Psychiatry* 71:4 (2008): 359–70, doi:10.1521/psyc.2008.71.4.359.

Goldberg, Stephen L., "Psychology's Contribution to Military Training," in *The Oxford Handbook of Military Psychology*, eds. Janice H. Laurence and Michael D. Matthews (New York: Oxford University Press, 2012), 241–61.

Gottman, John M., Julie S. Gottman, and Christopher L. Atkins, "The Comprehensive Soldier Fitness Program: Family Skills Component," *American Psychologist* 66:1 (2011): 52–57, doi:10.1037/a0021706.

Graham, Scott E., and Michael D. Matthews, eds., *Infantry Situation Awareness* (Arlington, VA: US Army Research Institute for the Behavioral and Social Sciences, 1999).

Grossman, Dave, *On Killing: The Psychological Cost of Learning to Kill in War and Society*, rev. ed. (New York: Back Bay, 2009), 5–17.

Hannah, Sean T., "The Authentic High-Impact Leader," in *Leadership Lessons from West Point*, ed. Doug Crandall (San Francisco: Jossey-Bass, 2007), 88–106.

Hill Jr., Randall W., James Belanich, H. Chad Lane, Mark Core, Melissa Dixon, Eric Forbell, Julia Kim, and John Hart, *Pedagogically Structured Game-Based Training: Development of the ELECT BiLAT simulation* (Marina Del Rey CA: University of Southern California Institute for Creative Technologies, 2006), http://www.dtic.mil/cgi-bin/GetTRDoc?AD=ADA461575.

Holsboer, Florian, and Marcus Ising, "Stress Hormone Regulation: Biological Role and Translation into Therapy," *Annual Review of Psychology* 61 (2010): 81–109.

Kahneman, Daniel, and Amos Tversky, "On the Reality of Cognitive Illusions," *Psychological Review* 103:3 (1996): 582–91, doi:10.1037/0033-295X.103.3.582.

Kalat, James W., *Biological Psychology*, 11th ed. (Belmont, CA: Wadsworth, 2012).

Kessler, Ronald C., Amanda Sonnega, Evelyn Bromet, Michael Hughes, and Christopher B. Nelson, "Posttraumatic Stress Disorder in the National Comorbidity Survey," *Archives of General Psychiatry* 52:12 (1995): 1048–60, doi:10.1001/archpsyc.1995.03950240066012.

Khashau, Ali S., Kathryn M. Abel, Roseanne McNamee, Marianne G. Pedersen, Roger T. Webb, Philip N. Baker, Louise C. Kenny, Preben Bo Mortensen, "Higher Risk of Offspring Schizophrenia Following Antenatal Maternal Exposure to Severe Adverse Life Events," *Archives of General Psychiatry* 65:2 (2008): 146–52, doi:10.1001/archgenpsychiatry.2007.20.

Klein, Gary, and Beth Crandall, *Recognition-Primed Decision Strategies, Research Note 96-36* (Alexandria, VA: US Army Research Institute for the Behavioral and Social Sciences, 1996).

Knapp, Deirdre J., and Tonia S. Heffner, *Validating Future Force Performance Measures (Army Class): End of Training Longitudinal Validation*, Technical Report 1257 (Arlington, VA: US Army Research Institute for the Behavioral and Social Sciences, 2009).

Kolditz, Thomas A., "Leading as if Your Life Depended on It," in *Leadership Lessons from West Point*, ed. Doug Crandall (San Francisco: Jossey-Bass, 2007), 160–87.

Kolditz, Thomas A., *In Extremis Leadership: Leading as if Your Life Depended on It* (San Francisco: Jossey-Bass, 2007).

Kyllonen, Patrick C., and Raymond E. Christal, "Cognitive Modeling of Learning Abilities: A Status Report of LAMP," in *Testing: Theoretical and Applied Issues*, ed. Ronna F. Dillon and James W. Pelligrino (New York: Freeman, 1989), 146–73.

Kyllonen, Patrick C., "Cognitive Abilities Testing: An Agenda for the 1990s," in *Personnel Selections and Classification*, ed. Michael G. Rumsey, Clinton B. Walker, and James H. Harris (Hillsdale, NJ: Erlbaum, 1994), 103–25.

Larson, Gerald E., Robyn M. Highfill-McRoy, and Stephanie Booth-Kewley, "Psychiatric Diagnoses in Historic and Contemporary Military Cohorts: Combat Deployment and the Healthy Warrior Effect," *American Journal of Epidemiology* 167:11 (2008): 1269–76, doi:10.1093/aje/kwn084.

Laurence, Janice H., and Michael D. Matthews, *The Oxford Handbook of Military Psychology* eds. (New York: Oxford University Press, 2012).

Laurence, Janice H. et al., *Effects of Military Experience on the Post-Service Lives of low-Aptitude Recruits: Project 100,000 and the ASVAB Misnorming*, Final Report 89-29 (Alexandria, VA: Human Resources Research Organization, 1989).

Lester, Paul B., P. D. Harms, Mitchel N. Herian, Dina V. Krasikova, and Sarah J. Beal, *The Comprehensive Soldier Fitness Program Evaluation, Report #3: Longitudinal Analysis of the Impact of Master Resilience Training on Self-Reported Resilience and Psychological Health Data* (Arlington, VA: Department of the Army, 2011).

Lester, Paul B., and Cynthia Pury, "What Leaders Should Know about Courage," in *Leadership in Dangerous Situations*, eds. Patrick J. Sweeney, Michael D. Matthews, and Paul B. Lester (Annapolis, MD: Naval Institute Press, 2011), 21–39.

Lukey, Brian J., and Victoria Tepe, *Biobehavioral Resilience to Stress* (Boca Raton, FL: CRC Press, 2008).

Lygre, Ragnhild B., and Jarle Eid, "In Search of Psychological Explanations of Terror," in *The Oxford Handbook of Military Psychology*, eds. Janice H. Laurence and Michael D. Matthews (New York: Oxford University Press, 2012), 114–28.

Maddi, Salvatore R., *Hardiness: Turning Stressful Circumstances into Resilient Growth* (New York: Springer, 2012).

Maddi, Salvatore R., Michael D. Matthews, Dennis R. Kelly, Brandilynn Villarreal, and Marina White, "The Role of Hardiness and Grit in Predicting Performance and Retention of USMA Cadets," *Military Psychology* 24:1 (2012), 19–28, doi:10.1080/08995605.2012.639672.

Matthews, Michael D., "Women in the Military: Comparison of Attitudes and Knowledge of Service Academy Cadets Versus Private College Students," in *Proceedings of the Psychology in the Department of Defense Thirteenth Symposium* (USAFA TR 92-2), ed. Anthony J. Aretz (Colorado Springs, CO: US Air Force Academy, Department of Behavioral Sciences and Leadership, 1992), 212–16.

Matthews, Michael D., "Comparison of US and Norwegian Cadets and College Students on Approval of Homosexuals Serving in the Military" (unpublished data, US Military Academy, West Point, NY).

Matthews, Michael D., "Where Eagles Soar: Positive Character and Success at West Point," paper presented at the Annual Meeting of the American Psychological Association, San Francisco, CA, August 2007.

Matthews, Michael D., "Toward a Positive Military Psychology," *Military Psychology* 20:4 (2008): 289–98, doi:10.1080/08995600802345246.

Matthews, Michael D., "Positive Psychology: Adaptation, Leadership, and Performance in Exceptional Circumstances," in *Performance Under Stress*, eds. Peter A. Hancock and James L. Szalma (Aldershot, UK: Ashgate, 2008), 163–80.

Matthews, Michael D., "Self-Reported Knowledge of Posttraumatic Stress Disorder and Posttraumatic Growth Among West Point Cadets" (unpublished raw data, United States Military Academy, West Point, NY, 2009).

Matthews, Michael D., "Character Strengths and Post-Adversity Growth in Combat Leaders," poster presented at the Annual Meeting of the American Psychological Association, Washington, DC, August 2011.

Matthews, Michael D., and Jarle Eid, "The Role of Women in the Military: An International Comparison," paper presented at the Annual Meeting of the American Psychological Association, Washington, D. C., August 2005.

Matthews, Michael D., Jarle Eid, Bjorn Helge Johnsen, and Ole C. Boe, "A Comparison of Expert Ratings and Self-Assessments of Situation Awareness During a Combat Fatigue Course," *Military Psychology* 23:2 (2011): 125–36, doi:10.1080/08995605.2011.550222.

Matthews, Michael D., Jarle Eid, Dennis R. Kelly, Jennifer K. S. Bailey, and Christopher Peterson, "Character Strengths and Virtues of Developing Military Leaders: An International Comparison," *Military Psychology* 18 suppl. (2006): S57–S68, doi:10.1207/s15327876mp1803s_5.

Matthews, Michael D., Morten G. Ender, Janice H. Laurence, and David E. Rohall, "Role of Group Affiliation and Gender on Attitudes Toward Women in the Military," *Military Psychology* 21:2 (2009): 241–51, doi:10.1080/08995600902768750.

Matthews, Michael D., Eddie C. Melton, and Charles N. Weaver, "Attitudes Toward Women's Roles in the Military as a Function of Gender and Race," in *Proceedings of the Psychology in the Department of Defense, Eleventh Symposium* (USAFA-TR-88-1), ed. Francis E. McIntire (Colorado Springs, CO: US Air Force Academy, Department of Behavioral Sciences and Leadership, 1988), 426–30.

McEwen, Bruce S., "Allostasis and Allostatic Load Implications for Neuropsychopharmacology," *Neuropsychopharmacology* 22:2 (2000): 108–24; doi:10.1016/S0893-133X(99)00129-3.

McFate, Montgomery, Britt Damon, and Robert Holliday, "What Do Commanders Really Want to Know? US Army Human Terrain System Lessons Learned from Iraq and Afghanistan," in *The Oxford Handbook of Military Psychology*, eds. Janice H. Laurence and Michael D. Matthews (New York: Oxford University Press, 2012), 92–113.

McGuigan, Frank J., *Experimental Psychology: A Methodological Approach* (Englewood Cliffs, NJ: Prentice Hall, 1968), 2–4.

Meili, Trisha, *I Am the Central Park Jogger: A Story of Hope and Possibility* (New York: Scribner, 2003).

Miller, Nita Lewis, Panagiotis Matsangas, and Aileen Kenney, "The Role of Sleep in the Military: Implications for Training and Operational Effectiveness," in *The Oxford Handbook of Military Psychology*, eds. Janice H. Laurence and Michael D. Matthews (New York: Oxford University Press, 2012), 262–81.

Miller, Nita Lewis, and Lawrence G. Shattuck, "Sleep Patterns of Young Men and Women Enrolled at the United States Military Academy: Results from Year 1 of a 4-Year Longitudinal Study," *Sleep* 28:7 (2005): 837–41.

Mortenson, Greg and David Oliver Relin, *Three Cups of Tea: One Man's Mission to Promote Peace... One School at a Time* (New York: Viking Penguin, 2006)

Ness, James, Denise Jablonski-Kaye, Isabell Obigt, and David A. Lam, "Understanding and Managing Stress," in *Leadership in Dangerous Situations*, eds. Patrick J. Sweeney, Michael D. Matthews, and Paul B. Lester (Annapolis, MD: Naval Institute Press, 2011), 40–59.

Norman, Donald A., *The Design of Everyday Things* (New York: Basic Books, 2002).

Ockenhouse, Christian F., Alan Magill, Dale Smith, and Wil Milhous, "History of US Military Contributions to the Study of Malaria," *Military Medicine* 170 (April Supplement 2005): 12–16, http://www.afids.org/AFIDS%20Milit%20Med%20Suppl%203-Malaria.pdf.

Parasuraman, Raja, "Neuroergonomics: Brain, Cognition, and Performance at Work," *Current Directions in Psychological Science* 20:3 (2011): 181–86, doi:10.1177/0963721411409176.

Pargament, Kenneth I., and Patrick J. Sweeney, "Building Spiritual Fitness in the Army: An Innovative Approach to a Vital Aspect of Human Development," *American Psychologist* 66:1 (2011): 58–64, doi:10.1037/a0021657.

Park, Nansook, "Congressional Medal of Honor Recipients: A Positive Psychology Perspective," paper presented at the Annual Meeting of the American Psychological Association, Washington, DC, August 2005.

Passer, Michael W., and Ronald E. Smith. *Psychology: The Science of Mind and Behavior* (New York: McGraw-Hill, 2009), 631–32.

Petty, Richard E., John T. Cacioppo, Alan J. Strathman, and Joseph R. Priester, "To Think or Not to Think: Exploring Two Routes to Persuasion," in *Persuasion: Psychological Insights and Perspectives*, ed. Timothy C. Brock and Melanie C. Green, 2nd ed. (Thousand Oaks, CA: Sage, 2005), 81–116.

Peterson, Christopher, Nansook Park, and Carl A. Castro, "Assessment for the US Army Comprehensive Soldier Fitness Program: The Global Assessment Tool," *American Psychologist* 66:1 (2011): 10–18, doi:10.1037/a0021658.

Peterson, Christopher, and Martin E. P. Seligman, *Character Strengths and Virtues: A Handbook and Classification* (New York: Oxford, 2004).

Pfeifer, Joseph W., and James Merlo, "The Decisive Moment: The Science of Decision Making under Stress," in *Leadership in Dangerous Situations*, eds. Patrick J. Sweeney, Michael D. Matthews, and Paul B. Lester (Annapolis, MD: Naval Institute Press, 2011), 230–48.

Pleban, Robert J., David E. Eakin, Marnie S. Salter, and Michael D. Matthews, *Training and Assessment of Decision-Making Skills in Virtual Environments*, Research Report 1767 (Alexandria, VA: US Army Research Institute for the Behavioral and Social Sciences, 2001).

Province, Charles M., *Patton's One-Minute Messages: Tactical Leadership Skills of Business Managers* (New York: Presidio, 1995), 65.

Reed, Brian, Chris Midberry, Raymond Ortiz, James Redding, and Jason Toole, "Morale: The Essential Intangible," in *Leadership in Dangerous Situations*, eds. Patrick J. Sweeney, Michael D. Matthews, and Paul B. Lester (Annapolis, MD: Naval Institute Press, 2011), 202–17.

Rohall, David E., Morten G. Ender, and Michael D. Matthews, "The Role of Military Affiliation, Gender, and Political Ideology on Attitudes Toward the Wars in Afghanistan and Iraq," *Armed Forces & Society* 33:1 (2006): 59–77.

Rumsey, Michael G., "Military Selection and Classification in the United States," in *The Oxford Handbook of Military Psychology*, eds. Janice H. Laurence and Michael D. Matthews (New York: Oxford University Press, 2012), 129–47.

Scales, Robert H., "Clausewitz and World War IV," *Military Psychology* 21 suppl. 1 (2009): S23–S35.

Seligman, Martin E. P., "Army Strong: Comprehensive Soldier Fitness," in *Flourish* (New York: Free Press, 2011), 126–51.

Seligman, Martin E. P., and Raymond D. Fowler, "Comprehensive Soldier Fitness and the Future of Psychology," *American Psychologist* 66:1 (2011): 82–86, doi:10.1037/a0021898.

Seligman, Martin E. P., and Michael Kahana, "Unpacking Intuition: A Conjecture," *Perspectives on Psychological Science: A Journal of the Association for Psychological Science* 4:4 (2009): 399–402, doi:10.1111/j.1745-6924.2009.01145.x

Seligman, Martin E. P., and Michael D. Matthews, eds., "Comprehensive Soldier Fitness." Special Issue, *American Psychologist* 66:1 (2001).

Seligman, Martin E. P., Tracy A. Steen, Nansook Park, and Christopher Peterson, "Positive Psychology Progress: Empirical Validation of Interventions," *American Psychologist* 60:5 (2005): 410–21, doi:10.1037/0003-066X.60.5.410.

Shay, Jonathan, *Achilles in Vietnam: Combat Trauma and the Undoing of Character* (New York: Simon & Schuster, 1994).

Sherif, Muzafer et al, *The Robbers Cave Experiment: Intergroup Conflict and Cooperation* (Middletown, CT: Wesleyan University Press, 1988).

Skinner, B. F., *Beyond Freedom and Dignity* (New York: Alfred A. Knopf, 1971).

Sledge, Eugene B., *With the Old Breed: At Peleliu and Okinawa* (New York: Presidio, 2007).

Smith III, Irving, *Why Black Officers Still Fail*, Strategy Research Project (Carlisle Barracks, PA: US Army War College, 2010), http://www.dtic.mil/cgi-bin/GetTRDoc?AD=ADA520126.

Stern, Fritz, "Einstein's Germany" in *E = Einstein: His Life, His Thought, and His Influence on Our Culture*, eds. Donald Goldsmith and Marcia Bartusiak (New York: Sterling, 2006), 97–118.

Sumner, Edwin M., *Sept–Oct 1943 The Cavalry Journal: General George Patton in Sicily* (Fort Riley, KS: The United States Cavalry Association, 1943).

Sweeney, Patrick J., "Trust: The Key to Combat Leadership," *Leadership Lessons from West Point*, ed. Doug Crandall (San Francisco: Jossey-Bass, 2007), 252–77.

Sweeney, Patrick J., Kurt T. Dirks, David C. Sundberg, and Paul B. Lester, "Trust: The Key to Leading When Lives Are on the Line," in *Leadership in Dangerous Situations*, eds. Patrick J. Sweeney, Michael D. Matthews, and Paul B. Lester (Annapolis, MD: Naval Institute Press, 2011), 163–81.

Sweeney, Patrick J., Michael D. Matthews, and Paul B. Lester, eds., *Leadership in Dangerous Situations* (Annapolis, MD: Naval Institute Press, 2011).

Sweeney, Patrick J., and Michael D. Matthews, "A Holistic Approach to Leading in Dangerous Situations," in *Leadership in Dangerous Situations*, eds. Patrick J. Sweeney, Michael D. Matthews, and Paul B. Lester (Annapolis, MD: Naval Institute Press, 2011), 373–91.

Sweeney, Patrick J., Vaida Thompson, and Hart Blanton, "Trust and Influence in Combat: An Interdependence Model," *Journal of Applied Social Psychology* 39:1 (2009): 235–64, doi:10.1111/j.1559-1816.2008.00437.x.

Tedeschi, Richard G., and Richard J. McNally, "Can We Facilitate Posttraumatic Growth in Combat Veterans?" *American Psychologist* 66:1 (2011): 19–24, doi:10.1037/a0021896.

Toomer, Jeffery K., *A Corps of Many Colors: The Evolution of the Minority Recruiting Effort at the United States Military Academy* (Research Report, Long Island University, 1997).

Tzu, Sun, *The Art of War* (New York: Delacorte Press, 1983).

US Department of the Army, *Army Leadership: Competent, Confident, and Agile*, Field Manual 6-22 (Washington, DC, 2006).

US Department of the Army, *Leadership Statements and Quotes*, Pamphlet 600–65 (Washington, DC, 1985).

US Department of the Army, *Staff Organization and Operations*, Field Manual 101-5 (Washington, DC, 1997), Chapter 5, http://www.au.af.mil/au/awc/awcgate/army/fm101-5_mdmp.pdf.

Varley, John, *The John Varley Reader: Thirty Years of Short Fiction* (New York: Berkeley, 2004), 271.

Weiss, Jeff, and Jonathan Hughes, "Implementing Strategies in Extreme Negotiations," *Harvard Business Review: Idea in Practice*, http://hbr.org/web/ideas-in-practice/implementing-strategies-in-extreme-negotiations.

Xiong, Glen L., and P. Murali Doraiswamy, "Does Meditation Enhance Cognition and Brain Plasticity?" *Annals of the New York Academy of Sciences* 1172:1 (2009): 63–69, doi:10.1196/annals.1393.002.

Zaccaro, Stephen J., Deanna Banks, Lee Kiechel-Koles, Cary Kemp, and Paige Bader, *Leader and Team Adaptation: The Influence and Development of Key Attributes and Processes*, Technical Report 1256 (Arlington, VA: US Army Research Institute for the Behavioral and Social Sciences, 2009).

Zaccaro, Stephen J., Eric J. Weiss, Rita M. Hilton, and Jack Jefferies, "Building Resilient Teams," in *Leadership in Dangerous Situations*, eds. Patrick J. Sweeney, Michael D. Matthews, and Paul B. Lester (Annapolis, MD: Naval Institute Press, 2011), 182–201.

Zajonc, Robert B., "Attitudinal Effects of Mere Exposure," *Journal of Personality and Social Psychology Monograph Supplement* 9:2, p2 (1968): 1–27, doi:10.1037/h0025848.

Zimbardo, Philip, *The Lucifer Effect* (New York: Random House, 2008).

INDEX

Abu Ghraib military prison, 47, 132, 224, 225
academic potential, assessing, 17
Achilles Hope and Possibility Five-Miler, 188
Achilles in Vietnam (Shay), 3
active duty soldiers, percent suffering from psychological trauma, 46
actors, hired to play indigenous people, 93, 102
adaptive killing, 184–187
adaptive thinking, 171
adaptivity, of future military leaders, 167
Adorno, Theodore, 168
adrenal cortex, 179
adversarial situations, 140, 141
adversity
    driving adaptation and innovation, 210
    extending to significant challenges, 75
    facing severe, 76
    leading to personal growth, 74
    resilience and personal growth and, 69
    resilience in the face of, 83
aerial flights, training for, 39
Afghanistan
    criminal actions of US military in, 224
    latest digital command and control technologies, 53
    military action in, 224
    no well-defined enemy in, 56
    PTSD rates associated with, 51
    stress-related disorders associated with, 44–45
    tactical command post, 149
African Americans
    believed to be mentally incapable of most military duties, 121–122
    competition to recruit for academies, 114
    excluded from the other officer corps in World War II, 108

integration into the military, 112, 118
    in the military, 112–114
    officers, 112, 113
    selecting army jobs with higher civilian transfer, 113
age, grit increasing with, 21
Agent Orange, exposure to, 70
aggression, 186
    resulting from punishment, 220
    social support and mediation shifting to and back, 91
    war formalizing and legitimizing use of violent, 213
aggressiveness, 140, 141
AHD (Army Human Dimension) doctrine, 125, 126, 127
air battles, simulated, 39
airborne training, 37
aircraft, 4, 13, 197
aircraft crews, training, 34
aircraft maintenance, as a group task, 116
Air Force Human Resources Laboratory (AFHRL), 18, 36
Air Force Officer Qualifying Test (AFOQT), 12
Air Force Officer Training School (OTS), 12, 13, 16, 159
Air Force Research Laboratory, 36
allostasis, concept of, 143
all-volunteer-force, replacing conscripted military, 80
American military personnel, largely culturally ignorant, 170
American Psychological Association (APA)
    division of military psychology, 3
    stance of in reexamining ethical standards, 226
    support of war effort in 1917, 13
Americans, believing themselves as superior, 90
"amplifiers," role in warfare, 7